4th International Conference on Preservation and Conservation Issues in Digital Printing and Digital Photography 2010

Journal of Physics: Conference Series Volume 231

London, United Kingdom
27-28 May 2010

Editors:

A. Fricker **P. Green**

ISBN: 978-1-61738-624-4
ISSN: 1742-6588

Printed from e-media with permission by:

Curran Associates, Inc.
57 Morehouse Lane
Red Hook, NY 12571

proceedings.com

Some format issues inherent in the e-media version may also appear in this print version.

Printed by Curran Associates, Inc. (2010)

For permission requests, please contact the Institute of Physics
at the address below.

Institute of Physics
Dirac House, Temple Back
Bristol BS1 6BE UK

Phone: 44 1 17 929 7481
Fax: 44 1 17 920 0979

techtracking@iop.org

Additional copies of this publication are available from:

Curran Associates, Inc.
57 Morehouse Lane
Red Hook, NY 12571 USA
Phone: 845-758-0400
Fax: 845-758-2634
Email: curran@proceedings.com
Web: www.proceedings.com

TABLE OF CONTENTS

Preface

These conference proceedings contain the written papers of the contributions presented at the 4th International Conference on: Preservation and Conservation Issues in Digital Printing and Digital Photography. The conference was held at the Institute of Physics, London, UK on 27th-28th May 2010. Previous conferences in this series took place in 2000, 2003 and 2006.

The aim of this conference series is to inform those responsible for the preservation of digitally printed materials about developments in digital photography and printing technologies. We aim to examine progress in research on inks and substrates and their significance for conservation and preservation issues and techniques. We also hope to develop links between related industries and the conservation/preservation world.

Research areas explored in this conference include current developments and future trends in digital printing and photographic technologies; the effect of environmental, storage and salvage conditions on the durability of digital prints and photographs; image processing techniques; image permanence considerations and standards for fastness, permanence and the role of scanning and file formats.

We would like to thank all participants for their contribution to the conference programme and these proceedings. Our thanks go to Ms C. Gu and Mr M. Sandy for chairing conference sessions. We are also grateful to Dawn Stewart and the Institute of Physics Conference Team for their invaluable support and assistance in arranging the conference and reception. Finally we would like to extend our thanks to the Society of Imaging Science and Technology (IS&T) for their sponsorship support.

The Editors

Acknowledgements

Conference Organising Committee: Ms A Fricker and Dr. P Green (London College of Communication, University of the Arts London).
Proceedings edited and compiled by Ms A Fricker and Dr. P Green.

Effects of Ozone on the Various Digital Print Technologies: Photographs and Documents

D Burge[1], N Gordeladze, J-L Bigourdan and D Nishimura

Image Permanence Institute at Rochester Institute of Technology, 70 Lomb Memorial Drive, Rochester, NY 14623, USA

E-mail: dmbpph@rit.edu

Abstract. The harmful effects of ozone on inkjet photographs have been well documented. This project expands on that research by performing ozone tests on a greater variety of digital prints including colour electrophotographic and dye sublimation. The sensitivities of these materials are compared to traditionally printed materials (black-and-white electrophotographic, colour photographic and offset lithographic) to determine if the digital prints require special care practices. In general, the digital prints were more sensitive to ozone than traditional prints. Dye inkjet prints were more sensitive to fade than pigment inkjet, though pigment was not immune. The dye sublimation, colour electrophotographic (dry and liquid toner), and traditional print systems were relatively resistant to ozone. Text-based documents were evaluated in addition to photographic images, since little work has been done to determine if the type of object (image or text) has an impact on its sensitivity to ozone. The results showed that documents can be more resistant to ozone than photographs even when created using the same printer and inks. It is recommended that cultural heritage institutions not expose their porous-coated, dye-based inkjet photos to open air for extended periods of time. Other inkjet prints should be monitored for early signs of change.

Introduction

The purpose of this project was to survey the most common digital print materials for their resistance to ozone. Digital prints already exist within cultural heritage collections and are expected to continually increase in quantity [1]. Deterioration of these materials in collections has already been reported [1]. Collection managers need a general overview of which digitally printed materials are sensitive to ozone and to what degree so that they can take measures to prevent decay as well as develop usage policies which restrict exposure. While much work has been done to examine the effects of ozone on individual or small numbers of digital print types [2], there has been no major survey that has incorporated the great variety of digital printing technologies, colorants, and papers. Additionally, previous work has focused on damage to pictorial images. This project includes examining the effects of ozone on text-based documents as well. There has also been little work to simultaneously examine digitally and traditionally printed materials to develop a context of risk for the modern materials. The audience for this report is collection care professionals at cultural heritage institutions; however, others, such as professional photographers and imaging manufacturers may find the results helpful.

[1] To whom any correspondence should be addressed.

Journal of Physics: Conference Series **231** (2010) 012001 doi:10.1088/1742-6596/231/1/012001

Methods

The test samples included inkjet, colour electrophotographic, dye sublimation, and digital press to represent digital prints. Chromogenic photo, black-and-white (B&W) electrophotographic, and offset lithography were used to represent traditional prints.

The inkjet prints were further sub-divided into dye and pigment, photo and document. The inkjet photo papers included cast-coat, porous photo-coated, polymer photo-coated, and fine art papers. Cast-coat papers are plain papers with a simple, porous ink-receiver layer applied to the surface (usually both sides). Porous-coated photo papers are resin-coated (also known as RC) papers with a more complex porous ink-receiver layer. These look and feel most like traditional colour photographs. Fine art inkjet papers use various types of high-quality artist's papers to which a porous ink-receiver layer has been applied. Porous-coated photo and fine art papers are usually coated on only one side of the paper. The porous coating on all these papers consists of mineral particles in a binder. The spaces between the particles form extremely small cavities that can absorb the inkjet ink during printing. Polymer-coated inkjet photo papers are resin-coated papers with an ink-receiver layer that swells when the ink is applied and then shrinks when the ink's solvent evaporates. The inkjet document papers used included both plain office and office papers treated specifically (either by chemical treatment or coating) to receive inkjet inks.

Plain papers and papers treated especially to receive electrophotographic toners (both black-and-white and colour) were used for the electrophotographic prints. Coated glossy print stock was used for the offset lithographic and all the digital press prints. Note that the digital presses were production-scale electrophotographic printers that use either dry or liquid toners. Multiple systems (printer/ink/paper combinations) were tested when possible. Table 1 shows the types of printers and papers tested as well as the number of systems (printer/ink/paper combinations) tested for each.

Table 1. Printers and papers tested.

Printer	*Paper*	*Purpose*	*No. of Systems Tested*
Inkjet – Dye	Porous Photo	Photo	3
Inkjet – Dye	Polymer Photo	Photo	3
Inkjet – Dye	Cast Coat	Photo	1
Inkjet – Dye	Plain Office	Document	3
Inkjet – Dye	Inkjet Office - Uncoated	Document	1
Inkjet – Dye	Inkjet Office - Coated	Document	1
Inkjet – Pigment	Porous Photo	Photo	2
Inkjet – Pigment	Fine Art Photo	Photo	3
Inkjet – Pigment	Plain Office	Document	3
B&W Electrophotographic	Plain Office	Document	3
B&W Electrophotographic	Laser	Document	1
Colour Electrophotographic	Plain Office	Document	3
Colour Electrophotographic	Colour Laser	Document	1
Dye Sublimation	Dye Sublimation	Photo	2
Digital Press – Dry Toner	Coated Glossy	Document	2
Digital Press – Liquid Toner	Coated Glossy	Document	1
Chromogenic Silver-Halide	Chromogenic Silver-Halide	Photo	2
Offset Lithography	Coated Glossy	Document	1

Colour test targets for each system were printed in duplicate. The targets consisted of cyan, magenta, and yellow patches in ten approximately equal intervals of density from minimum density

Journal of Physics: Conference Series **231** (2010) 012001 doi:10.1088/1742-6596/231/1/012001

(Dmin) to maximum density (Dmax), neutral patches with 20 intervals, as well as two non-printed patches (Dmin). Text targets consisted of black text on a white field and white text on a black field. The font was New Times Roman and the text ranged in size from 8 point to 14 point. "Best Photo" and "Photo Enhanced" printer settings were selected when available for photo-specific printing systems. Default settings were used for document printing systems. After printing, all samples were left to dry at 21°C and 50% RH in the dark for two weeks before testing. All targets were read using a Gretag Spectrolino/Spectroscan (no UV filter, 2° observer, D50 illuminant) for CIELAB L*a*b* both before and after ozone exposure. Delta E values were then calculated. Results for each printer and paper type were averaged to determine that technology's sensitivity to ozone. Text targets were assessed visually to determine the smallest readable font after ozone exposure.

The ozone-exposure chamber used was custom built for IPI by Codori Enterprises. The ozone was produced by means of an ultra-violet lamp. The samples were exposed at 5 ppm ± 0.25 ppm for 2 weeks. The temperature and humidity within the chamber were held constant at 25°C ± 2°C and 50% RH ± 5% RH.

Results

Table 2 shows the average delta E values for the maximum cyan, magenta, and yellow density patches for each printer and paper type. Table 3 shows the average delta E values for a mid-tone neutral patch (approximately 70% of Dmax so as to include the three primary colours plus black inks) and the maximum density neutral patch for each printer and paper type. While most grey tones in colour digital prints are made up of cyan, magenta, yellow, and black colorants together, the maximum density black is often black colorant only. Additionally, some inkjet systems use dyes for the cyan, magenta, and yellow colours and a pigment for black.

Table 2. Effect of ozone on delta E of cyan, magenta, and yellow colorants.

Printer	*Paper*	*Cyan*	*Magenta*	*Yellow*
Inkjet – Dye	Porous Photo	33	62	44
Inkjet – Dye	Polymer Photo	6	2	1
Inkjet – Dye	Cast Coat	46	96	73
Inkjet – Dye	Plain Office	8	7	4
Inkjet – Dye	Inkjet Office - Uncoated	9	6	5
Inkjet – Dye	Inkjet Office - Coated	25	57	51
Inkjet – Pigment	Porous Photo	12	5	3
Inkjet – Pigment	Fine Art Photo	10	8	1
Inkjet – Pigment	Plain Office	8	7	2
B&W Electrophotographic	Plain Office	NA[a]	NA[a]	NA[a]
B&W Electrophotographic	Laser Office	NA[a]	NA[a]	NA[a]
Colour Electrophotographic	Plain Office	4	1	2
Colour Electrophotographic	Colour Laser Office	8	2	2
Dye Sublimation	Dye Sublimation	2	3	2
Digital Press – Dry Toner	Coated Glossy	3	3	1
Digital Press – Liquid Toner	Coated Glossy	8	7	1
Chromogenic	Chromogenic	1	0	1
Offset Lithography	Coated Glossy	4	3	1

[a] NA = not applicable. B&W electrophotographic does not have cyan, magenta, and yellow toners, only black.

Journal of Physics: Conference Series **231** (2010) 012001 doi:10.1088/1742-6596/231/1/012001

Table 3. Effect of ozone on delta E of mid-tone and maximum black.

Printer	*Paper*	*Mid-tone*	*Dmax*
Inkjet – Dye	Porous Photo	48	36
Inkjet – Dye	Polymer Photo	4	1
Inkjet – Dye	Cast Coat	56	65
Inkjet – Dye	Plain Office	3	1
Inkjet – Dye	Inkjet Office - Uncoated	5	1
Inkjet – Dye	Inkjet Office - Coated	14	1
Inkjet – Pigment	Porous Photo	5	1
Inkjet – Pigment	Fine Art Photo	3	0
Inkjet – Pigment	Plain Office	2	1
B&W Electrophotographic	Plain Office	1	1
B&W Electrophotographic	Laser	1	0
Colour Electrophotographic	Plain Office	2	2
Colour Electrophotographic	Colour Laser	4	5
Dye Sublimation	Dye Sublimation	3	2
Digital Press – Dry Toner	Coated Glossy	1	2
Digital Press – Liquid Toner	Coated Glossy	3	1
Chromogenic Silver-Halide	Chromogenic Silver-Halide	0	0
Offset Lithography	Coated Glossy	1	0

Both ink and paper selections have a significant effect on the overall sensitivity of the system as a whole (printer/ink/paper combination). The porous-coated photo paper, cast-coated paper, and coated-office paper printed with dye inkjet were all more likely to fade than those printed with dye inkjet using polymer or plain office papers or any paper printed with pigment inks. While the pigment inkjet prints were less sensitive to damage by ozone than dye inkjet, they were not immune. The electrophotographic and dye sublimation as well as the traditional print types were more resistant to ozone-induced fade than dye or pigment inkjet. In several cases, the maximum density neutral patches faded at a slower rate than mid-tone neutral patches suggesting that text and line art that uses full density black may be more resistant to ozone-induced decay over time. This applied to both dye and pigment inkjet with the exception of dye inkjet on porous photo- or cast-coated paper. The text targets were all readable at all font sizes for all print technologies after exposure to ozone even though some of the inkjet prints were severely faded.

Table 4 shows the average delta E values for each technology for the density minimum (white) areas of the print.

Table 4. Effect of ozone on delta E of white areas of the print.

Printer	*Paper*	*Average ΔE*
Inkjet – Dye	Porous Photo	1
Inkjet – Dye	Polymer Photo	0
Inkjet – Dye	Cast Coat	4
Inkjet – Dye	Plain Office	1
Inkjet – Dye	Inkjet Office - Uncoated	1
Inkjet – Dye	Inkjet Office - Coated	3
Inkjet – Pigment	Porous Photo	1
Inkjet – Pigment	Fine Art Photo	1
Inkjet – Pigment	Plain Office	1

B&W Electrophotographic	Plain Office	1
B&W Electrophotographic	Laser	1
Colour Electrophotographic	Plain Office	1
Colour Electrophotographic	Colour Laser	1
Dye Sublimation	Dye Sublimation	1
Digital Press – Dry Toner	Coated Glossy	3
Digital Press – Liquid Toner	Coated Glossy	3
Chromogenic Silver-Halide	Chromogenic Silver-Halide	0
Offset Lithography	Coated Glossy	5

Visually, none of the papers discoloured severely, but several did yellow slightly, most noticeably the digital press and offset lithographic prints. The cast-coat inkjet and coated-inkjet office papers also yellowed.

In addition to fading and yellowing, one of the porous photo-coated papers also showed cracking and delamination of the ink-receiver layer. While rare, this is more objectionable because a faded print may still be readable/viewable, but a print that has lost some or its entire ink-receiver layer would be considered a total loss.

Previous work by the field had already shown that dye-based inkjet printed on porous-coated photo papers could be highly sensitive to ozone attack. That result was replicated here; however, it was unexpected that the dye inkjet on plain paper would be so much less sensitive to ozone exposure. It had been assumed that with plain papers, ozone would be able to attack the image from both sides making it more sensitive to ozone. It is not clear why this did not happen. Since the prints on the plain and porous photo papers in this test were made using different printers, a second experiment to compare prints made with the same printers was performed to validate the effect. The methodology was the same as above including the 5 ppm for two weeks exposure. Three printers were tested using both porous photo paper and plain office paper. Table 4 shows the delta E values for neutral mid-tone patches printed on both photo and plain papers on each of three dye inkjet printers.

Table 5. Delta E values of mid-tone neutral on photo and plain papers.

Printer	*Photo Paper*	*Plain paper*
Printer 1	31	2
Printer 2	57	2
Printer 3	66	2

In each case, the inks on plain paper were more resistant to ozone-induced fade than the same inks on porous photo paper.

Conclusions

As a result of the research the following conclusions were drawn:

- Exposing inkjet photographs to ozone can cause fading of colorants, yellowing of print papers, and cracking of ink-receiver coatings.
- Both dye and pigment inkjet printed materials are more sensitive to ozone-induced deterioration than electrophotographic, dye sublimation, or traditionally printed materials.
- Dye inkjet prints on porous-coated papers (photo, cast, fine art, and coated inkjet office) are significantly more sensitive to ozone-induced fade than any other type of printer/ink/paper combination.
- Inkjet documents on plain papers can be more resistant to ozone than inkjet photographs.
- Exposure to ozone can cause yellowing of digital press and offset lithographic print papers.

It is therefore recommended that cultural heritage institutions not expose their porous-coated, dye-based inkjet photos to open air for extended periods of time. Use of inert, plastic enclosures during handling is highly advised. Pigment-based inkjet photos, dye-based inkjet on polymer photo paper, and all inkjet printed documents are all more resistant to fade than dye inkjet on porous paper but they should be monitored over time for early signs of decay. Dye sublimation photographs and all electrophotographic prints may be treated using the same care practices that are currently being used for traditionally printed materials.

References

[1] Burge D, Nishimura D and Estrada M 2009 Summary of the DP3 Project Survey of Digital Print Experience within Libraries, Archives, and Museums *Archiving 2009*, (Arlington, VA, USA: Imaging Science & Technology) pp 133-136

[2] Mason P and Carreira L 2007 Image Permanence: Comparing the Technologies *NIP23: International Conference on Digital Printing Technologies and Digital Fabrication* (Anchorage, Alaska, USA: Imaging Science & Technology) p 732

Acknowledgments

The authors would like to acknowledge the support of the Andrew W. Mellon Foundation, who provided the funding for this project as part of a larger grant-funded research program to examine digital print preservation issues. We would also like to thank the Printing Applications Laboratory at the Rochester Institute of Technology for providing the digital press and offset print samples. In addition, Harman Technology Inc. and the Eastman Kodak Company provided printers and papers to support the project. The Image Permanence Institute is jointly sponsored by the Society for Imaging Science and Technology and the Rochester Institute of Technology.

Primary Author Biography

Daniel M. Burge, Senior Research Scientist, has been a full-time member of the Image Permanence Institute (IPI) staff for the last 19 years. He received his B.S. degree in Imaging and Photographic Technology from the Rochester Institute of Technology in 1991. He managed IPI's enclosure testing services from 1991 to 2004. In 2004, he took over responsibility for all of IPI's corporate-sponsored research projects. Currently he is investigating digital print stability and storage issues.

Journal of Physics: Conference Series **231** (2010) 012002 doi:10.1088/1742-6596/231/1/012002

Effect of short-term simulated daylight exposure on optically brightened papers

Phil Green and Mike Honess

London College of Communication, Elephant and Castle, London SE1 6SB

pj.green@lcc.arts.ac.uk

Abstract. Two optically brightened commercial papers were subjected to 72 hours continuous exposure to graphic arts daylight simulating fluorescent lamps at 1.4 and 4.4 klux. At the higher illuminance level, a CIELAB colour difference of 1.0 was reached in approximately 3 and 50 hours for the two papers. The colour differences produced by the 1.4 klux exposure were greater than that predicted by assuming a linear relationship between intensity and colour change, by a factor of 1.8.

1. Introduction

Papermakers achieve high whiteness and brightness levels through the addition of fluorescent whitening agents (FWAs) which modify reflection from a surface by emitting visible light on excitation by photons with frequencies in the ultra-violet. The majority of commercially-available papers are optically brightened in this way.

A concern with FWAs is their stability over time, especially when irradiated with light which includes UV. Quenching of fluorescence leads to a visible colour change, which is typically a reduction in whiteness and brightness attributes accompanied by an apparent yellowing.

Test methods for investigating the effect of light exposure on prints and photographs are based on accelerated fading tests using high-intensity illumination, typically in the range 30-80 thousand lux (klux) [1]. These can be combined with exposures at lower intensities in order to determine the extent of reciprocity failures and estimate the effect of exposures in normal viewing conditions.

Of interest in the present study is the extent of fading of the optically brightened papers used in commercial printing, when displayed under graphic arts viewing conditions. For conservation purposes it is common to exclude UV from the illumination source, but the FWAs in commercial white papers require an element of UV radiation in the source to achieve the intended brightness and whiteness levels, and this is incorporated in the specification of the reference illuminant in ISO 3664:2009 [2].

This paper is part of a series of investigations into colour changes in fluorescent papers after printing which include accelerated exposure conditions, but in this study we chose to use actual graphic arts viewing booths at a typical illumination level, and at an augmented level achieved by fitting additional lamps in the booth.

2. Experimental setup

Two papers with high levels of optical brightener were selected. These were a commercial gloss art paper with 10% recycled fibre content, and a premium bond paper. They were exposed in a Verivide colour viewing cabinet for 72 hours to fluorescent lamps simulating a D50 illuminant, having been selected from the middle of previously unopened packets. The lamp specifications given by the manufacturer [3] conform to the requirements of ISO 3664:2009 with a Colour Rendering Index [4] of 98, and the lamp is also stated to conform to the metamerism index specifications of ISO 3664:2009 in both visible and UV regions.

Two papers were exposed with 4 lamps, giving an illuminance on the paper surface of 4426 lux. One of the papers was also exposed for the same time period to two lamps, giving 1433 lux; the illuminance being lower than might be expected since the two lamps were further from the exposure area and at a more glancing angle.

The lab where the test was conducted has a stable temperature and humidity, but these were not recorded during the test.

Measurements of reflected radiance were made in-situ at 20-minute intervals over a 72-hour period using a Jeti Specbos 1200 telespectroradiometer. Measurements were recorded at wavelengths 380-780nm at 1nm intervals and the in-situ radiance data were converted to XYZ.

Measurements were also made of a white PTFE plaque of 10mm thickness at the same position as the paper samples at the beginning and end of each exposure and measurement run. The reflectance of the tile was used to normalise the in-situ tristimulus measurements of the papers to Y=100 for a perfect reflecting diffuser. The tristimulus values of the illuminant were computed in the same way, and the paper and illuminant tristimulus values used to compute CIELAB coordinates and colour differences relative to the first measurement in each run.

The short-term reproducibility of the measurements was evaluated by 10 repeat measurements of the PTFE plaque. The standard deviation under the 1433 and 4426 klux exposures was respectively 0.05% and 0.48% of the measurement.

3. Results

The resulting colour differences are shown in Figures 1-3. It can be seen that in each case there is a relatively rapid initial change, followed by a slower continuing change. In the case of the 1433 lux exposure, the colour change is not significant after approximately 12 hours, while at 4426 lux the change continues throughout the 72 hours of exposure. The approximate time to 1 CIELAB ΔE^*_{ab} colour change is 3 hours for paper A and 50 hours for paper B.

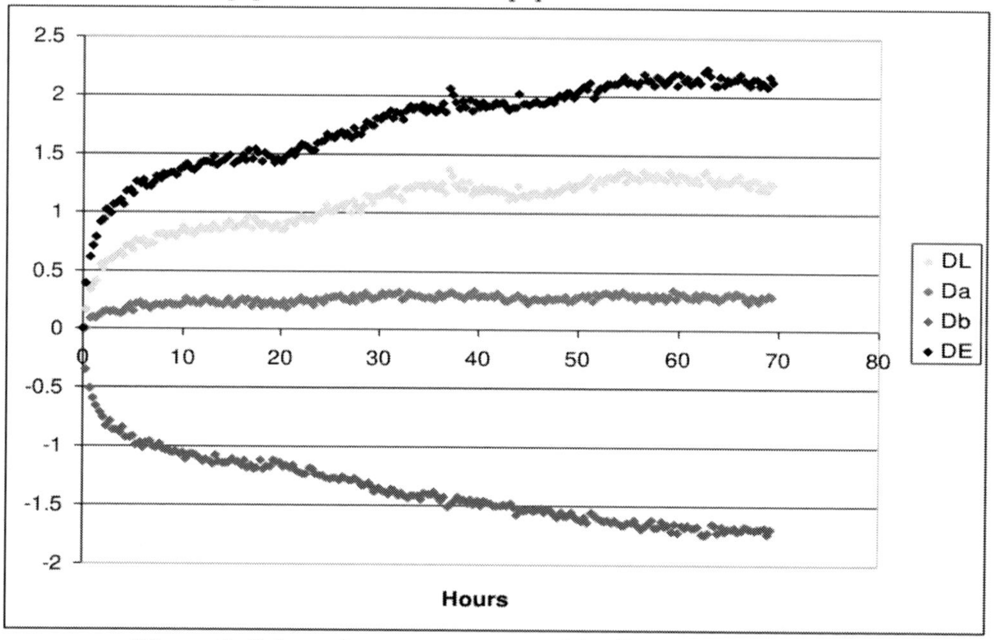

Figure 1. Colour changes on commercial offset gloss at 4426 lux

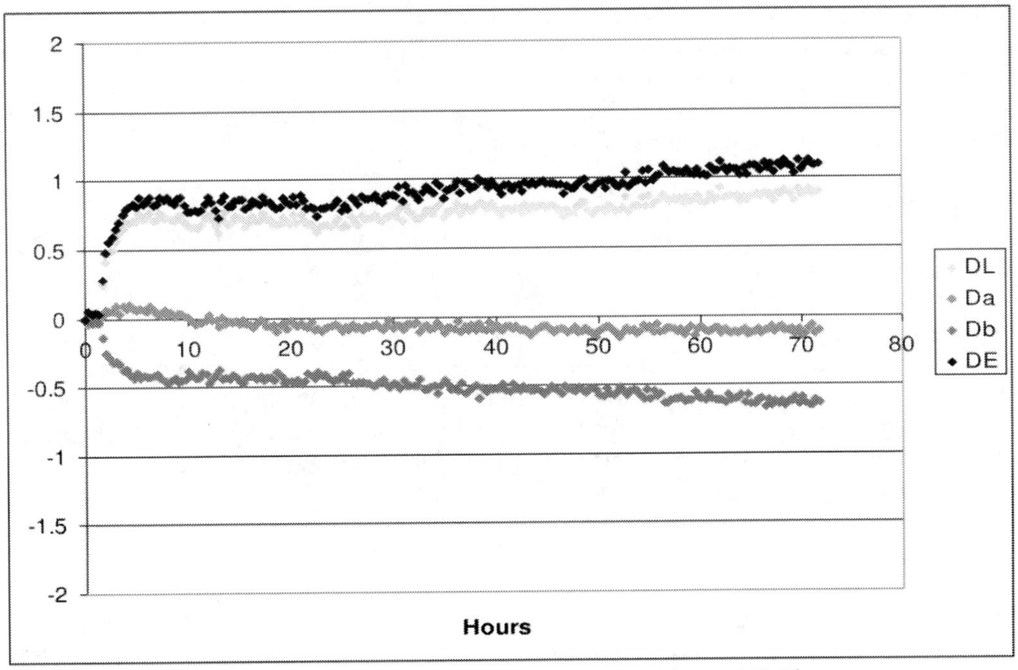

Figure 2. Colour changes on premium inkjet at 4426 lux

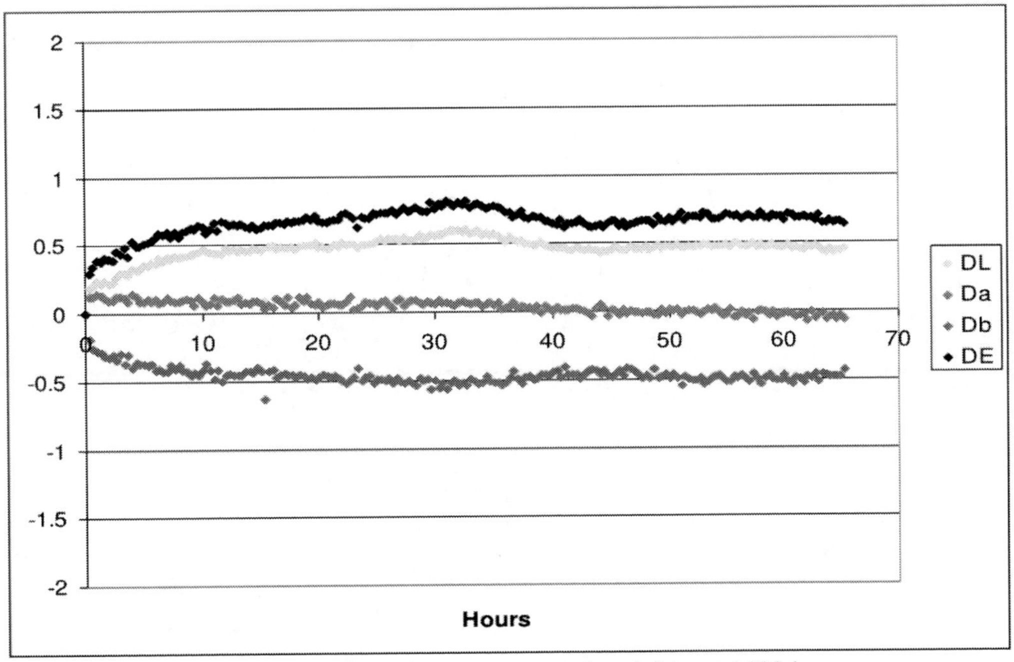

Figure 3. Colour changes on premium inkjet at 1433 lux

On the commercial offset paper the largest changes in the L*, a* and b* components are in b*, which is consistent with quenching of FWAs. Interestingly, it can be seen in Figures 2 and 3 that the proportion of colour change which is accounted for by the b* component is greater at the lower intensity, and the change in b* is almost constant between 1433 and 4426 lux.

It can be seen in Figure 3 that approximately 50% of the change in b* has occurred in the first 20 minutes of exposure at 1433 lux.

Comparing the results at 1433 and 4426 lux, the lower level of illuminance produces a much larger change than would be predicted by assuming that the exposure effects of 10klux for one hour is equivalent to 1klux for 10 hours; the change at the end of the 72 hour exposure was 1.8 times what would be expected if the colour change were linearly related to the exposure intensity.

It is assumed that the FWA quenching is permanent and that it would not recover after a period of keeping in dark conditions, but this was not tested.

4. Conclusions

Non-accelerated exposures to graphic arts viewing illumination lead to significant colour changes in optically-brightened papers over relatively short time periods. The time to reach a 1 CIELAB ΔE^*_{ab} colour change under 4426 lux varied greatly between the two papers, suggesting that fading characteristics of papers should be known in order to select suitable papers for different applications.

Since the magnitude of change in the b* component of colour difference, which is most associated with radiative changes to FWAs, is fairly consistent between the two exposure levels it is possible that there is a threshold level at which change occurs, and it would be interesting to investigate this further. The rapid initial change in this b* component at 1433 lux also suggests a need to investigate further at lower illuminances and exposure periods.

The magnitude of change under the lower exposure intensity was substantially greater than predicted by assumptions of reciprocity.

References

[1] ISO 18937:2010 (WD) Imaging materials — Photographic reflection prints — Methods for measuring indoor light stability, ISO, Geneva
[2] ISO 3664:2009 Graphic technology and photography — Viewing conditions, ISO, Geneva
[3] Manufacturer's data sheet for Philips TL950 lamp
[4] CIE Publication 13.3, Method of Measuring and Specifying Colour Rendering Properties of Light Sources, CIE, Vienna (1995)

Journal of Physics: Conference Series **231** (2010) 012003 doi:10.1088/1742-6596/231/1/012003

Developments in the recovery of colour in fine art prints using spatial image processing

A Rizzi[1] and C Parraman[2]
[1]Dip. Tecnologie dell'Informazione
Università degli Studi di Milano, 26013, Crema (CR), Italy
[2]Centre for Fine Print Research,
University of the West of England, Bristol BS3 2JT, UK

[1]Alessandro.Rizzi @unimi.it [2]Carinna.Parraman@uwe.ac.uk

Abstract. Printmakers have at their disposal a wide range of colour printing processes. The majority of artists will utilise high quality materials with the expectation that the best materials and pigments will ensure image permanence. However, as many artists have experienced, this is not always the case. Inks, papers and materials can deteriorate over time. For artists and conservators who need to restore colour or tone to a print could benefit from the assistance of spatial colour enhancement tools. This paper studies two collections from the same edition of fine art prints that were made in 1991. The first edition has been kept in an archive and not exposed to light. The second edition has been framed and exposed to light for about 18 years. Previous experiments using colour enhancement methods [9,10] have involved a series of photographs that had been taken under poor or extreme lighting conditions, fine art works, scanned works. There are a range of colour enhancement methods: Retinex, RSR, ACE, Histogram Equalisation, Auto Levels, which are described in this paper. In this paper we will concentrate on the ACE algorithm and use a range of parameters to process the printed images and describe these results.

1. Introduction

Artists have a wide range of materials from which to choose, most of which are produced to a high standard. Commercial manufacturers of, for example, pigments, papers and inks will disclose product information on the permanence of their materials such as a star rating for oil and water-colours, or a Wilhelm rating for inkjet materials. In some cases there is an assumption that pigment colourants will last longer than a dye, but due to highly sophisticated manufacturing methods and specialist requirements for colourants these boundaries between organic and inorganic are beginning to be blurred. However, these materials – dyes, inks, papers, film-stock, digital data, are subject to change and deterioration and the best approach to preservation is to limit the circumstances where deterioration does occur.

But what of the materials and images that have already begun to deteriorate? Artists and conservators who need to restore images to their original condition would benefit from the assistance of colour enhancement tools to reproduce colours that aim to be as close to the original as possible. In many cases, an original is not always at hand for comparison, which presents a problem for the restorer to guess what was the original colour. For this research study, we have had access to editions of both exposed and unexposed fine art prints. The exposed prints are works that have been on display since 1991, under glass and in the same north-east facing position. The unexposed prints, which are from the same edition, have been stored in a print archive at the Centre for Fine Print Research. These two editions present a very useful opportunity for the restorer to compare firstly how the exposed edition has deteriorated and secondly when applying digital image enhancement methods to a scanned image of the exposed images what colour information can be restored.

In order to obtain a good quality image or a restored image, the process of adjusting images can be a time consuming and a costly procedure. In this paper, we consider the use of unsupervised colour enhancement methods as final automatic processing or pre-processing for printing. We tested some classic widely used algorithms such as Histogram Equalisation and Auto Levels and compared them to alternative approaches that utilise *Spatial Colour Algorithms* (SCA). The term *Spatial Colour Algorithm* refers to a family of algorithms that re-compute wavelength/energy arrays into calculated colour appearance arrays, or preferred color enhancement arrays, according to the spatial distribution of pixel values in the scene [1]. Their basic idea is to mimic the behaviour of the human vision system. Among these algorithms, we describe the following: Retinex [2-4], RSR [5], ACE [6]. SCAs are generally parameter dependent [1]. We do not intend to consider fine tuning for each of the techniques, rather to consider an average parameter set for each one and then test if this approach can go in the user preference direction. Previous preliminary tests of this approach on display and prints are reported in [7-10]. In previous sample tests (using photographs and artworks) we have used the full range of enhancement methods. As we now have an understanding of their behaviour and results, for the purpose of this paper we have tested using just the ACE algorithm. As ACE can be regarded as a representative of this family, we have used a range of parameters for filtering (see section 5, table 1).

2. Typical problems and scenarios

The following sections describe typical problems that have been considered and addressed and scenarios where enhancement methods have been found to be useful and an outline of the enhancement algorithms.

2.1. Film restoration

Wet-process film materials are the result of a chemically unstable process, subject to fading over time. This fading is irreversible and in several cases photochemical restoration of faded prints is problematic and not always possible. In these cases, digital colour restoration can address the problem. Faded film images are dull, have poor saturation and an overall colour cast. This is due to the bleaching of one or two chromatic layers of the film. Since it is necessary to deal with lost chromatic information, restoring the colour of faded movies is more complex than a simple colour balance.

Digital techniques allows easily multiple restoring solutions, but still the restoration process can be a very delicate and long process, based mainly on the professional personal skills of the restoring technicians. The restoration market demands for automatic solutions, but the big diversity of all the possible film conditions makes every automatism difficult to implement.

2.2. Image enhancement for the printing of digital photographic images

User objectives for digital capture have more far reaching objectives: documentation, collection of images for recording, uploading to Internet, dissemination of images, prints and artworks through the Internet, recording of events and preparing images in preparation for printing eg. wedding or holiday albums, the process of selecting images therefore reflects these requirements.

2.3. Limited devices or non standard external conditions

As more spontaneous images are captured by phone, events may be captured in an environment that is under poor or artificial lighting conditions. The intention is to consider a range of photographic conditions and reasons for image capture, such as artefacts that are poorly lit (possibly due to museum or conservation issues), conditions where flash photography is prohibited or unsuitable, scenes taken in artificial light conditions, de-saturated scenes, atmospheric influences, photography of flat artworks, objects taken against a white background and underwater photographs are sampled.

2.4 Image enhancement for drawings, artworks and digitally generated images
As artists, designers and archivists require scanning and photography of artworks and artefacts, these might be used for digital archives, for the Internet, promotional materials. Some might be flat scanned, such as maps or small prints and photographs, and some might be very large that it is difficult to obtain a uniform lit surface. These artworks may be textured or collaged, digitally generated or manipulated artworks, black and white prints, commercial CMYK prints, newspaper images, scanned artwork, colour photographs, black and white photographs.

3. Description of colour enhancement methods

In order to address a range of problem images, we have worked with two categories of image enhancement methods. The first are classic global enhancement methods that are included in photo editing software which enables the user to adjust brightness, saturation, contrast by hand and in digital imaging software such as Photoshop: histogram equalisation, auto levels and auto-colour. This global approach maps every pixel using the same value.

The second is a spatial image enhancement approach (SCA) [1], which is inspired by human visual system characteristics and share both global and local colour correction characteristics. This family of algorithms makes a pixel-by-pixel comparison to re-compute the pixels based on the context of the scene. Considered examples are Retinex, RSR, ACE. The main characteristics of these methods are described here.

3.1. Histogram equalisation and curves
In commercial software applications, such as Photoshop, there is a range of tools for the fine-tuning of an image. Therefore, improving the dynamic range of an image can be undertaken using the histogram in Levels by changing the input and output levels; or through Curves by adjusting any point on a 0-255 tonal scale; or by assigning target values to the highlight and shadow pixels using either the Levels of Curves.

Photoshop automatic tools process the image on a global level, therefore making an assessment of an image by building a histogram, and applying corrections to the whole image. As demonstrated in figure 1, by stretching the black and white points, a brighter image is obtained. In the lower image, the histogram shows peaks in the black and white and nothing in between. However, using auto levels to an image that has high contrast no improvement is made to the mid tones (figure 1 and 2).

3.2. Auto levels
Auto Levels defines the lightest and darkest pixels in each colour channel as white and black and then stretches the in- between pixel values proportionately. Because Auto Levels adjusts each colour channel individually, it may remove or introduce colour casts.

3.3. Auto colour
Takes an average distribution from the darkest, midtone to the lightest areas of an image, by neutralising the midtones and clipping the white and black pixels. This can be undertaken in Curves on a more precise level by assigning up to 14 points on a curve, or by using the auto levels. Because each channel is adjusted individually, the algorithm may remove or introduce colour casts.

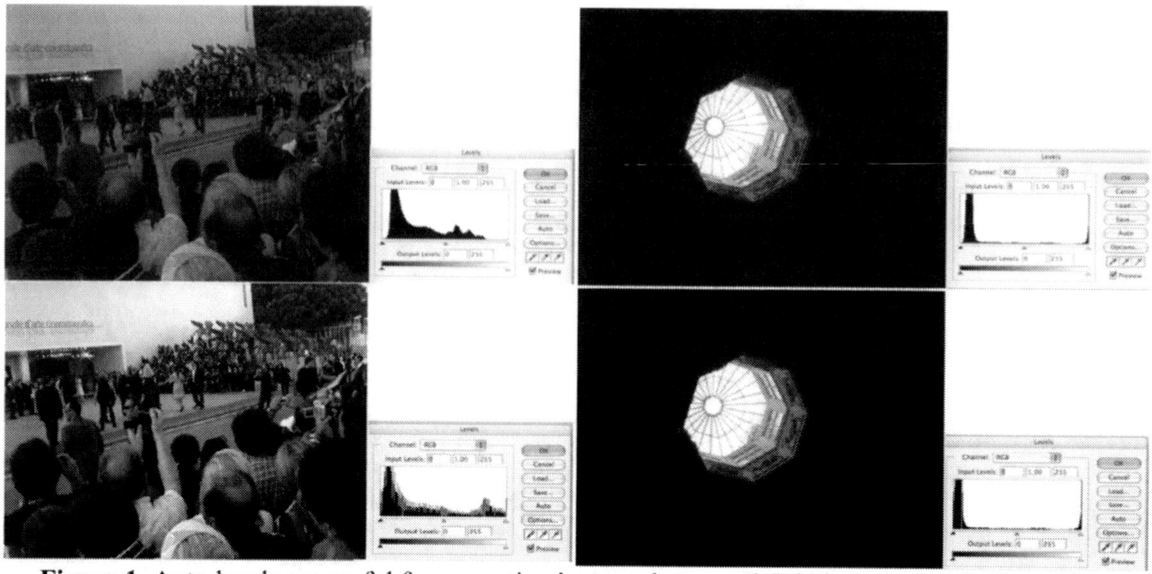

Figure 1. Auto levels are useful for correcting images that are slightly imbalanced (left), unlike figure 2, which benefits from spatial tuning parameters

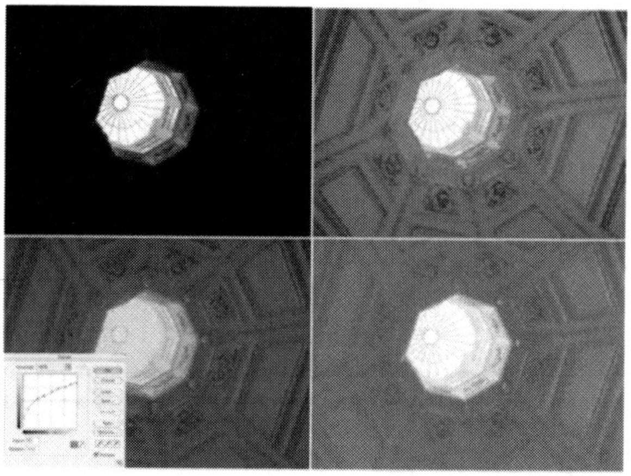

Figure 2. Image sampled using Auto colour command in Photoshop (top left) and Retinex (top right), hand correction using curves (bottom left), ACE Slope=5 (bottom right)

3.4. Retinex
Based on the human colour perception system, the term is a composite of 'retina' and 'cortex'. The Retinex algorithm, mimicking human vision, computes image colour appearance, and is described by Edwin Land in 1971 [2]. It performs global and local filtering according to the way it scans the input image. Several implementations have been developed so far [4].

3.5. RSR
Random Spray Retinex, (RSR) [5] is one of the implementations of the original Retinex model. In this implementation, locality is performed by random sprays of a target point rather than paths.

3.5. ACE
ACE [6] is an algorithm for unsupervised enhancement of digital images. The implementation of ACE

follows the classic SCA scheme: the first stage re-computes each pixel according to the content and the relative position of the values in the image while the second stage map the computed pixel values into the available device range.

4. Background to the Miniature Print Portfolio

This collection of prints, made in 1991, demonstrates a range of traditional (etching, stone litho) and more contemporary print methods from that time (screenprint, 4-colour-offset litho, laser-print, inkjet). In the early 1990s many artists were using photomechanical approaches to reproduce, drawings and paintings. More contemporary processes such as laser printing and inkjet were also being employed. The miniature print portfolio comprises staff from the CFPR, the School of Creative Arts, and MA students on the MA multidisciplinary printmaking course. Each artist produces an edition of 60 prints. The image must conform to 7.6 x 10.2cm and printed on a 20 x 25cm sheet of mould made paper, which is then presented in portrait format. These are the only requirements or limitations. The Centre now has an archive of mini prints from 23 years. It represents a collection of tens of thousands of individual prints, produced by hundreds of staff, students and invited artists using every print process. Prints have been exhibited nationally and internationally, in galleries and print shows from China to South America.

From the total of 60 prints in this edition, only a few of the prints have faded. It is interesting to note that the images printed using the more commercial/industrial processes, which employ process colours, have changed. As new processes are brought into the market, then artists are keen to try new materials and processes. However as demonstrated here not all these materials guarantee longevity. Due to the use of fugitive inks, some of the contemporary printing methods (4-colour-offset litho, laser-print, inkjet) that include process colours, in particular the magenta and yellow colours, have considerably faded. It was considered to be impossible to restore these as no actual colour residue remained on the print. The following section will highlight a few of the faded prints and compare the unexposed to the exposed.

4.1 Description of the prints

The following examples are the more extreme casualties of exposure to light. A digital scan has been made of both the unexposed print and the exposed print and compared here.

"Poppy Field" (figure 3). An early CMY inkjet print onto a lightweight 70gsm paper, cut around the image and tipped onto a fine art paper. The yellow and magenta inks have completely disappeared from the exposed image; only the cyan ink layer remains.

"Abstract" (figure 4). A screenprint printed with pigment water-based inks onto sections of gold leaf. Due to tighter health and safety legislation in art colleges in the 90s, water-based screenprinting inks were beginning to replace the traditional solvent based screen-inks. This ink-series was produced by TW Graphics Company, that have a high pigment content and dry to a mat finish. In the original image a speckled layer of white is printed over the main section revealing the darker colours and gold through the gaps. In the exposed image the over-layer of white and the dark under-layer of pink has faded, which has resulted in the blue layer and the gold appearing darker and more visible.

"Dune: (figure 5). An offset photo-litho printed from a zinc plate onto Somerset cream mould made paper, printed in black. The exposed image and the colour of the paper have faded slightly.

"Still Life" (figure 6). A four-colour half-toned offset litho print. In the exposed image both the yellow and magenta layers have faded.

"Starbird" (figure 7). A full colour laser print, the image is cut and tipped onto a fine art paper. In the exposed image the yellow has faded and the other colours appear de-saturated.

"Whitebait" (figure 8). A screenprint with Daler Rowney water based inks onto cream Somerset mould made paper. The white pigment has faded, which is particularly apparent around the top of the bag. The colour of the paper has also faded from cream to white, which gives the appearance of the grey-blue background appearing darker in the exposed image.

Figure 3. "Poppy field"

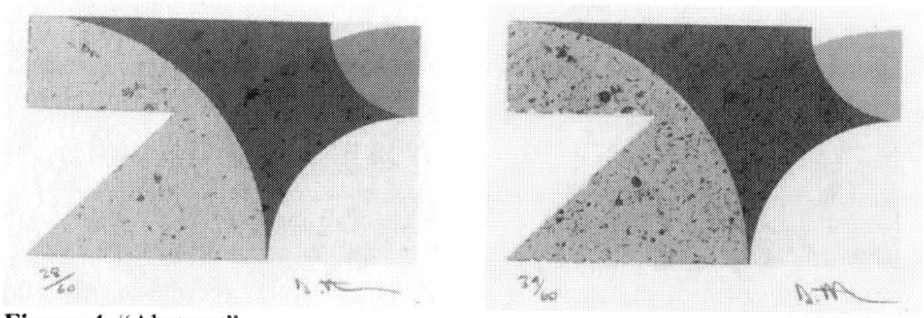

Figure 4. "Abstract"

Figure 5. "Dune"

Figure 6. "Still Life"

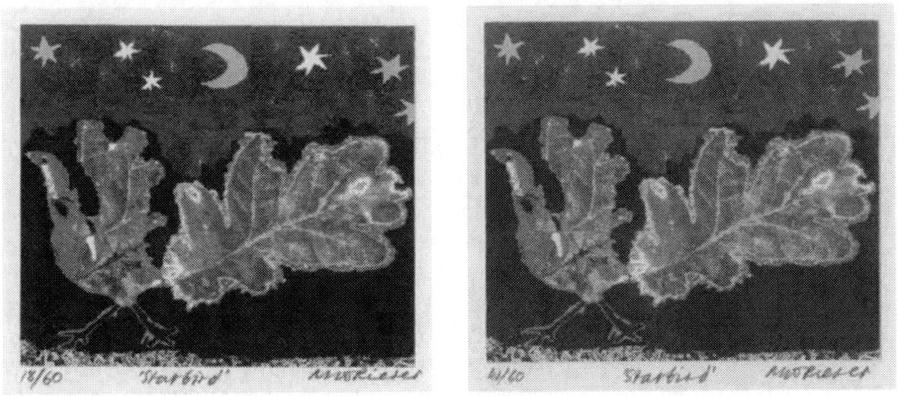

Figure 7. "Starbird"

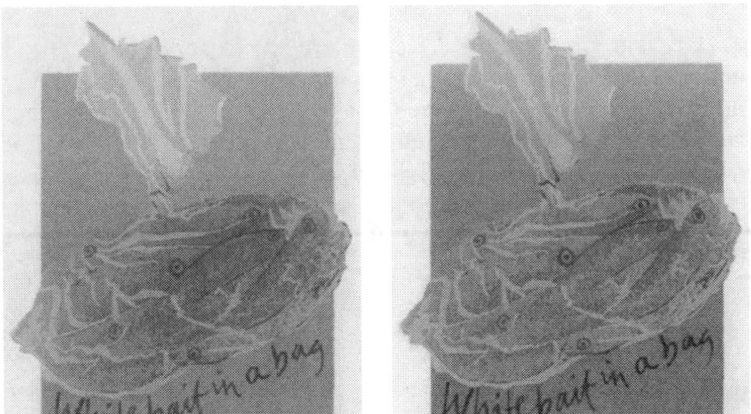

Figure 8. "Whitebait"

5. Test setup and results

All the faded images were scanned on an Epson RGB scanner and saved as .tiff formats. These were then converted to 24 bit image files in preparation for processing using the ACE algorithm.

Due to the lack of colour channels in the most faded prints, we expected that the prints would yield limited results. Furthermore, we have chosen the most deteriorated images and therefore are presented as a worst case scenario. Based on previous experience of how the algorithms performed we used the ACE algorithm on these prints. The parameters we used for these tests were:

Table 1. ACE Parameters

Parameter	Parameter range(s)
Subset selection	Sub-sampled at a sub-factor of LL2
Chromatic comparison	SL5 SL20 Sign
Distance	0.01

The following figure 9 illustrates the faded prints that have been subjected to the ACE parameters. They are arranged so that the faded image appears in the top left corner. In all of the cases the prints cannot be described as being close to the original unfaded image. However, there were some surprising results. Through the application of the spatial algorithm to the image "Poppy field" in which only one of the three colours remained, some traces of the yellow and a deepening of the blue to create the appearance of a fuller range of colours. If some colour information is still present within a chromatic channel, SCA are able to amplify it in a global and local way producing a pleasant result. From the user point of view there is no need to choose or select the areas on which to apply the filtering. In the monochrome images as exampled in "Dune", contrast and dynamic range is improved. For the "Still life" image, although only two channels remained, the algorithm was able to achieve a range of tones, find and enhance ghost colours and deepen the existing colours. Similar results were achieved on the "Abstract" and "Whitebait". Although the processed images appear much darker than their original unfaded versions, the algorithm was able to increase a range of tones and colour information. Other prints such as the "Leafbird", the dark blue background became lighter and the texture of the paper was also exaggerated on the mould made papers.

6. Conclusion

In general it was found that since ACE is a Gray World driven algorithm, in some images it can enhance particular chromatic dominances, in this case, colour can significantly be changed and it can result in a better image but with an extreme colour shift. Fine-tuning of the parameters in order to obtain these results requires some knowledge of the strengths and weaknesses of each of the algorithms. It was generally found that for the majority of black and white images the spatial correction methods are useful in increasing the tonal range. Significantly, it has been demonstrated that the spatial image enhancement methods are able to achieve surprising results in regenerating an image from a single channel to the appearance of three.

Figure 9. The variations of processed images using ACE Sign, Slope 5 and Slope 20

References

[1] Rizzi, A. and McCann, J.J. "On the behavior of spatial models of color", *IS&T/SPIE Electronic Imaging 2007,* S. Jose (California – USA), 28 January – 1 February 2007.

[2] Land, E.and McCann, J.J. ``Lightness and Retinex Theory'', J. Opt. Soc. Am. {\bf 61}, 1-11 (1971).

[3] Marini, D. and Rizzi A.``A computational approach to color adaptation effects'', Im. and Vis. Comp. **18** (13), 1005-1014 (2000).

[4] Provenzi, E., Rizzi, A. De Carli, L., Marini, D. "Mathematical definition and analysis of the Retinex algorithm" *Journal of Optical Society of America A*, Vol. 22, December (2005).

[5] Provenzi, E., Fierro W., Rizzi, A. De Carli, L., Gadia, D., Marini, D. "Random Spray Retinex: a new Retinex implementation to investigate the local properties of the model" *IEEE Transactions on Image Processing*, Vol. 16, Issue 1, pp. 162-171, January 2007.

[6] Rizzi, A., Gatta, C., Marini, D. "A New Algorithm for Unsupervised Global and Local Color Correction", *Pattern Recognition Letters*, Vol 24 (11), pp. 1663-1677, July 2003.

[7] Rizzi, A., Gatta, C., Taraschi, F., Zanardini, M., Abbiati, M. "Un algoritmo per la valutazione percettiva delle interfacce visuali", *DDD Disegno e Design Digitale*, Vol. 7 (2), Apr/Jun 2003, ed. Poli-Design (Politecnico di Milano).

[8] Rizzi, A., Gatta, C., Maggiore, M., Agnelli, E., Ferrari, D., Negri, D. "Automatic Lightness and Color Adjustment of Visual Interfaces", *HCI-Italy 2003,* Torino (Italy), October 2003.

[9] Parraman, C. and Rizzi, A. "User Preferences in Color Enhancement Unsupervised Methods for Printing", *IS&T/SPIE Electronic Imaging 2007,* S.Josè (California – USA), 28 January –1 February 2007.

[10] Parraman, C. and Rizzi, A. "Searching User Preferences in Printing: A Proposal for an Automatic Solution", *Printing Technology SpB06*, St Peterburg (Russia), June 2006.

[11] Rizzi, A., Gatta, C., Slanzi, C., Ciocca, G., Schettini, R. Unsupervised Color Film Restoration Using Adaptive Color Equalization, *Lecture Notes in Computer Science*, Volume 3736, Dec 2006, Pages 1 – 12.

[12] Rizzi, A. Perceptual colour film restoration, Preservation and Conservation Issues Related to Digital Printing and Digital Photography, Institute of Physics, London, 24th - 25th April 2006.

Journal of Physics: Conference Series **231** (2010) 012004 doi:10.1088/1742-6596/231/1/012004

Robust watermarking on copyright protection of digital originals

C Gu and X Y Hu

College of packaging & printing engineering, Tianjin University of Science & Technology, Tianjin, 300222, China

E-mail: guchong527@gmail.com, emma_huxy@yahoo.com.cn

Abstract. The issues about the difference between digital vector originals and raster originals were discussed. A new algorithm based on displacing vertices to realize the embedding and extracting of digital watermarking in vector data was proposed after that. The results showed that the watermark produced by the method is resistant against translation, scaling, rotation, additive random noise; it is also resistant, to some extent, against cropping. This paper also modified the DCT raster image watermarking algorithm, using a bitmap image as watermark embedded into target images, instead of some meaningless serial numbers or simple symbols. The embedding and extraction part of these two digital watermark systems achieved with software. Experiments proved that both algorithms are not only imperceptible, but also have strong resistance against the common attracts, which can prove the copyright more effectively.

1. Introduction

With the rapid development of computers and Internet, different kinds of digital originals have become quite popular. Meanwhile, the copy and transformation of them turned much easy that the traditional copyright protection is not enough any more. How to protect the digital originals' copyright is a very pressing problem.

As a branch of Information Hiding technology, digital watermarking is a possible approach to counter abuses of digital multimedia data, such as texts, audio data, images, movies. Digital watermarking adds a structure called watermark to the target data object imperceptibly to the users and inseparably from the object, to identify the copyright and protect the legal right of the author.

Watermarking must have strong invisibility and robustness, because of the object here is both commodity and intellectual production. Digital watermark can be many forms, like images, words, numerals, etc. it shouldn't affect the use of digital productions. However, when the dispute happens, the information encoded in the watermark can be used to identify the copyright owner or to detect tampering.

2. Digital watermarking for different digital originals

Two-dimensional prepress digital originals can be classified into either raster or vector digital data. Raster data represents an image composed by a 2D array of pixels. This kind of files is related to the resolution, so its quality can be easy depredated by rotation, scaling, and other geometrical transformations. Vector data files employ geometrical primitives such as points, lines, and polygons to represent objects using mathematics equations. Unlike the raster files, the vector digital files are able to be scaled and rotated without loss of quality. Due to the difference between raster data and vector data, used algorithms of watermarking must be different.

However, the watermark used in the copyright protection of both vector and raster originals should meet the following basic requests:

- Imperceptibility: the embedded watermark must be invisible in the target originals;
- Robustness: the watermarking originals can resilient against some common changes: enlarge/shrink, rotated, object insertion/deletion, additive random noise, cropping, and so forth.
- Vindicability: the watermark can be extracted significantly and used for the copyright certification.

3. Watermarking algorithm of vector originals

3.1. Embedding

In vector graphics, almost all entities are defined by some attributes related to coordinates, so it is possible that the watermark can be embedded by the vertex coordinates. Meanwhile, the points, lines and other objects are usually located all over the file, so the embedded watermark information is hidden dispersedly in space; which can reduce the instability caused by the concentration of embedding.

In order to protect the original vertices, all the embedding experiments were realized by drifting coordinates and adding a new attribute. Find the original entities and drift the vertex coordinates first, and then add a unique attribute for each new vertex, so the watermark can be written in it later. The testing file in experiments is a DXF format digital map; figure 1 shows what the map looks like.

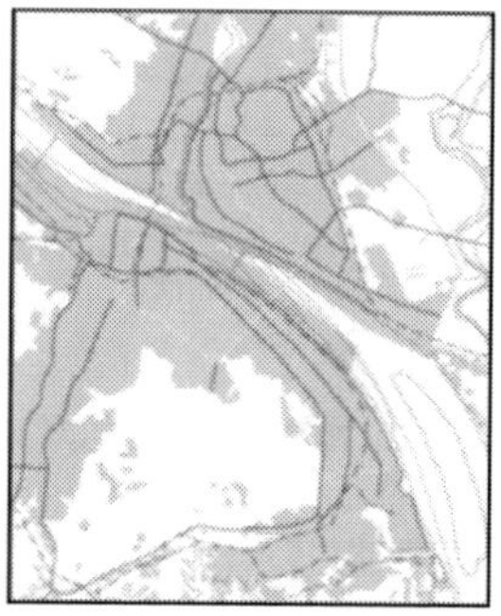

Figure 1. DXF Digital map in experiments.

The watermark embedding process is as follows, also shown in figure 2.

Step1, Search the file structure, find the locations of every stored coordinate. In DXF files, the coordinates are saved after a particular pair of group code and group values;

Step2, Read the *x, y* coordinates (since the discussed objects are 2D vector files, so all the *z* coordinates of the geometric entities in the data is always 0);

Step3, Reset the locations of the points within the precision allowed to modify, and rewrite the points into the file structure;

Step4, Add copyright watermark into the new coordinates, the watermark in experiment is "TUST Copyright";

Step5, Set the drift-points to "invisibility" to avoid the visual impact of graphics files.

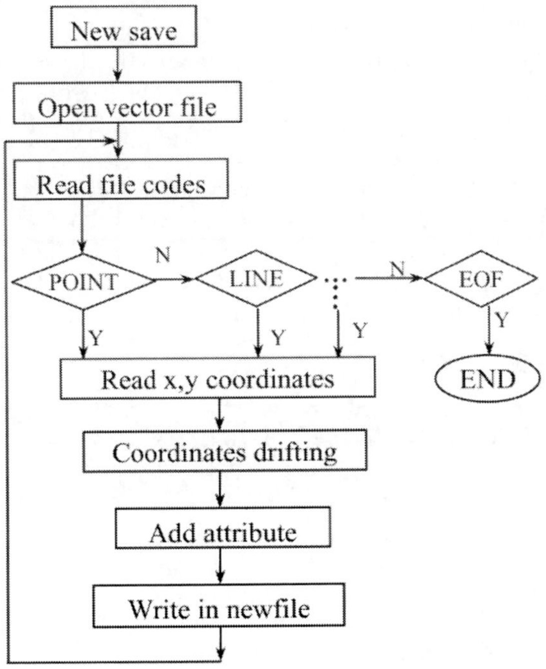

Figure 2. Flow of watermark embedding.

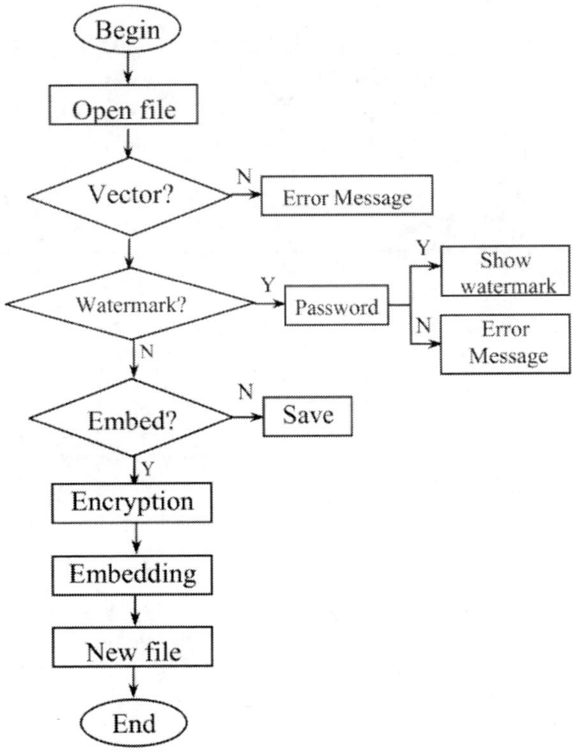

Figure 3. Flow of realization of watermark system.

The amount of embedded watermark is increased with the entities number contained in the vector graphics; this can be seen from table 1. The more entities in the file, the more locations of embedded watermark will be found, and the better dispersion the watermarks have. In this way, the effects of malicious tampering or damaging attacks on the watermark would be much smaller.

Table 1. Numbers of embedded watermark.

Entity numbers	Data-density (coordinate numbers)	Watermark capacity (bit numbers)
372	310	57
682	596	57
1463	1237	87
2541	2074	101
4095	3296	240
6482	5309	360
8543	7018	480
10605	8980	540

3.2. Detection

Watermark detection is executed once opening the graphic file in the program. When the program opens a file, data structure will be read line by line, if we find the unique embedded watermark in the file, the password will be asked to enter, if the entered password is correct, watermark information will be displayed; otherwise the file can not be opened. The detection flow is shown in figure 3.

Journal of Physics: Conference Series **231** (2010) 012004 doi:10.1088/1742-6596/231/1/012004

3.3. Experiments and Results

3.3.1. *Perceptibility*. Because all the drift-point coordinates are added "invisibility" attribute artificially, that means the locations that watermarks written in can not be seen, then embedded watermarks are totally invisible. Therefore, if the files with watermark be opened in graphics software such as AutoCAD, Illustrator, CorelDraw, etc, the watermark will not reduce the visual quality at all.

3.3.2. *Resiliency against attacks*. The watermark embedding is built on the basis of coordinates, so translation and scaling do not have any influence on the watermark. Other robustness evaluations of the embedded watermark are tested after various attacks, and then use the VC++ program to test. Table 2 shows the results of the experiment, in which the cropping test is detecting the cropping maps down to 1/16~1/2., the rates given in the table is the average rate of different methods.

Table 2. Error rates due to various attacks.

	Translation	Scaling	Rotation	Additive random noise		Cropping			
				α=50 cm	α=80 cm	1/16	1/8	1/4	1/2
Error rate (%)	0	0	0	2.1	4.9	9.1	6.4	5.7	0

Experiments showed that the watermarks produced by the method are resilient against shift, scaling and rotation. When random noise having the amplitudes of either $\alpha = 50$ cm or $\alpha = 80$ cm is added to the vertex coordinate, the error rate of this algorithm is under 5%; the accuracy of watermark system under cropping is also within the acceptability. Therefore, if drift-point number reaches to a certain quantity, the robustness against cropping is going to be stronger.

4. Watermarking algorithm of raster originals

Raster images are made of many picture elements that we call pixels. When the pixels are small enough, and the distance between two of them short enough, people can not be able to tell. In this way, we can take advantage of human visual redundancy to embed watermarks, or embed watermarks into the position where human eye can not easily detect. The algorithm used in this paper is based on Discrete Cosine Transform (DCT) method of transform domain algorithm.

The host image in experiments is a 480×480 pixel grayscale image of Lena, which saved as tiff format, shown in figure 4, and the watermark image is school badge image of Tianjin university of science & technology, saved as Bmp format, shown in figure 5.

Figure 4. Host image.

Figure 5. Watermark image.

4.1. Embedding

The embedding starts with the area division step, which tries to produce 8×8 pixel arrays in the host image, since the height of 480 is divisible by 8, there will be 60 arrays along the height of the image and there will be also 60 arrays along the width of the image. Then each array will be transformed with DCT.

According to the study, Human Vision System is relatively sensitive to low frequency DCT coefficients, and high frequency part is fragile to attacks, which will affect the robustness of watermark. Therefore, the watermark should be embedded into the intermediate frequency DCT coefficients of the host image. The colored part in figure 6 shows the arrays represent the intermediate frequency DCT coefficients.

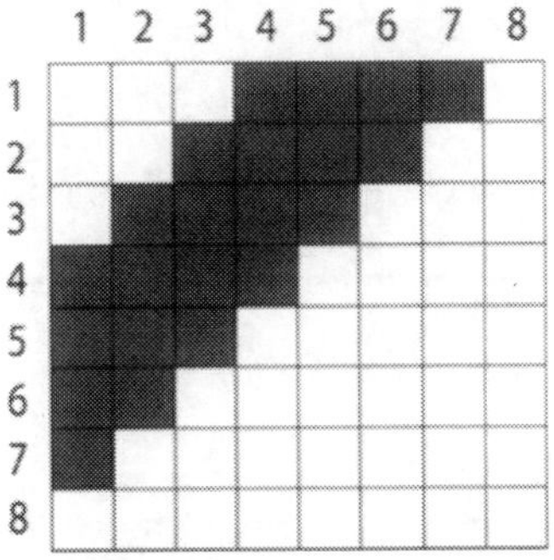

Figure 6. An 8×8 DCT coefficient array.

The level of variance reflects the smoothing degree of the divided arrays; the watermark image should be embedded into the position which has the largest variance here. So we calculate the average gray level m and variance of the arrays δ^2 by equation1 and equation 2.

$$m = \frac{1}{n^2}\sum_{i=0}^{n-1}\sum_{j=0}^{n-1} x(i,j) \tag{1}$$

$$\delta^2 = \frac{1}{n^2}\sum_{i=0}^{n-1}\sum_{j=0}^{n-1}[x(i,j)-m]^2 \tag{2}$$

Where n represents the number of row pixels, x is the gray value of each pixel. After computing, the binary watermark image was transformed into one-dimensional row vector.

To add random sequence generated from the row vector into the intermediate frequency DCT coefficients of n arrays that have the highest δ^2, DCT coefficients must be calculated for each array, and the arrays need to undergo an Inverse Discrete Cosine Transform (IDCT) in order to be written back as a image, shown in figure 7.

Figure 7. Watermarked image.

Figure 8. Extracted watermark.

4.2. Extraction

Prior to extraction, the host image and the watermarked image should be compared in DCT domain. By means of checking whether the DCT coefficients used to embed the watermark, the watermark can be reconstructed. According to the variance of the arrays, where the DCT is applied to the watermarked image can be determined.

After that, one-dimensional watermark sequence was generated by the order of sequence, as well as the complexity of the arrays, and reconstructs the watermark image by the watermark sequence, and makes it as an identification of the copyright.

4.3. Experiments and results

4.3.1. *Perceptibility*. We first tested perceptibility of the watermark using 15 volunteers, all non-expert in watermarking. We asked each one of them of she/he could tell a watermarked image if they are presented with a pair of same size images printed on a piece of paper, one watermarked and the other not. None could tell the watermarked image from the non-watermarked image.

In order to observe the image quality of watermarked image objectively, the PSNR (Peak Signal to Noise Ratio) value of the image is calculated using the equation 3 and if the PSNR value is greater than 35dB, the watermarked image is within acceptable degradation levels.

$$PSNR = 10 \times \lg((2^n - 1)^2 / MSE) \tag{3}$$

Where n means the number of bits per sample value, the MSE represents mean square error between the host image and the watermarked image. The PSNR value obtained for this test case was 44.6032 dB. This is well above 35 dB thus implying that the watermarked image is within acceptable degradation levels, and an appropriate image quality of the image that is being watermarked is maintained.

4.3.2. *Resiliency against attacks*. Watermark's robustness is based on whether the watermark can be extracted after experiencing a variety of attacks. In this paper, we chose several common attacks to test, such as modified, stained, cropping, format changing, printing and scanning. The similarity value calculated by equation 4 was used for determine the difference between the original watermark and extracted watermarks after attacking.

$$sim = \frac{W * \hat{W}}{\sqrt{W * W}} \tag{4}$$

Where $\hat{W}$ represents detected watermark; W represents original watermark; sim means similarity value。

When the similarity value is over 0.7, we can identify that the watermark experiment is successful. Table 3 shows the results of the experiments under various attacks.

According to table 3, the similarity value between watermark without modification and the original watermark was 0.9275, and similarity values of other extracted watermarks from various attacked images are all over 0.8. The results showed that this algorithm can identify the copyright definitely.

Table 3. Similarity of various attacks

	No modification	7% modified	12% cropping	9% stained	reformatting	printing & scanning
Similarity	0.9275	0.8487	0.8206	0.8096	0.9021	0.8017

5. Conclusion

In this paper, we analyzed the characteristics of graphics and images, studied their data properties, and found two kinds of visual redundancy location, and finally proposed two different watermarking algorithms for both two kinds of file. The experimental results determined the feasibility of the methods that could be the foundation for the further research in this field.

References

[1] Zhao Y 2006 *Sequence watermarking robust to random geometrical attack*(China: Chinese

Journal of Electrinics) pp634-637

[2] Ohbuchi R et al.1999 *Proc. Int. Conf. on Computer Graphics*(Canada: Canmore) pp170-177

[3] Ohbuchi R, Ueda H and Endoh S 2002 *Proc. Int. Conf. on Multimedia and Expo* (Switzerland: Lausanne) pp1-4

[4] Ohbuchi R, Ueda H and Endoh S 2003 *Proc. Int. Conf. on the Shape Modeling*(Korea: Seoul)pp1-10

[5] Schyndel van R G, Tirkel A Z and Osborne C F 1994 *Proc. Int. Conf.on Image Processing* vol2 pp86-90

[6] Bender W, Gruhl D, Morimoto N 1995 *Techniques for data hiding*: *Proc. of the SPIE* (CA:San Jose) *Storage and Retrieval for Image and Vedio Database* vol3 pp2420-2425

Journal of Physics: Conference Series **231** (2010) 012005 doi:10.1088/1742-6596/231/1/012005

Freeze-Drying Wet Digital Prints: An Option for Salvage?

M C Jürgens[1] **and N Schempp**[2]

[1] Photograph Conservator, Schumacherstr. 98, 22767 Hamburg, Germany
[2] Schempp Bestandserhaltung GmbH, Max-Planck-Str. 12, 70806 Kornwestheim, Germany

E-mail: post@martinjuergens.net

Abstract. On the occasion of the collapse of the Historical Archive of the City of Cologne in March 2009 and the ensuing salvage effort, questions were raised about the use of freeze-drying for soaked digital prints, a technique that has not yet been evaluated for these materials. This study examines the effects of immersion, air-drying, drying in a blotter stack, freezing and freeze-drying on 35 samples of major digital printing processes. The samples were examined visually before, during and after testing; evaluation of the results was qualitative. Results show that some prints were already damaged by immersion alone (e.g. bleeding inks and soluble coatings) to the extent that the subsequent choice of drying method made no significant difference any more. For those samples that did survive immersion, air-drying proved to be crucial for water-sensitive prints, since any contact with the wet surface caused serious damage. Less water-sensitive prints showed no damage throughout the entire procedure, regardless of drying method. Some prints on coated media suffered from minor surface disruption up to total delamination of the surface coating due to the formation of ice crystals during shock-freezing. With few exceptions, freeze-drying did not cause additional damage to any of the prints that hadn't already been damaged by freezing. It became clear that an understanding of the process and materials is important for choosing an appropriate drying method.

1. Introduction

On March 3, 2009, the six-story building of the Historical Archive of the City of Cologne collapsed within minutes into a large crater created by the faulty construction of a subway line beneath the street. During the ensuing six-month initial salvage effort of physically damaged, humid or soaked photographs, documents, maps, illuminated manuscripts and books, the question was raised whether wet digitally printed matter could be vacuum freeze-dried along with other archival materials or if it needed to be sorted out and treated separately. An investigation was initiated to determine the effects of freeze-drying on samples of the most common digital printing processes.

When facing large numbers of moist or wet archival materials, speed is an essential factor. In order to avoid excessive bleeding of inks, dissolution of water-sensitive coatings and mould growth, prints must either be dried quickly and safely, or, if this is not possible, frozen. Freezing instantly stops deterioration that occurs in wet materials and it gives the salvage team time to consider a variety of drying methods. While much is known about salvage methods for traditionally printed documents and photographic materials, only few studies have been published so far on the sensitivities of digital prints to liquid water and on methods of drying wet prints. These studies indicate that prints of

Journal of Physics: Conference Series **231** (2010) 012005 doi:10.1088/1742-6596/231/1/012005

different processes show different sensitivities to water: while dye diffusion thermal transfer prints survive flooding quite well, for example, since they do not tend to wet at all, inkjet prints with swellable (polymer) coatings absorb large amounts of water and become tacky to the touch or even suffer from total dissolution of the ink receptor layer. It has been shown that microporous inkjet prints, on the other hand, are generally less sensitive to water damage [1].

However, available investigations have not examined the option of freeze-drying, a method that is common for drying large numbers of documents or books within a short period of time in order to avoid ink bleeding, physical distortion, blocking and microbial activity. In vacuum freeze-drying, water is removed from the documents while these remain frozen. Within a vacuum, the frozen water held in the documents sublimates, leaves the document and is reformed as a solid on a condensing surface in a separate chamber. As drying progresses, the temperature within the vacuum chamber is incrementally increased in order to promote the release of residual water bound in the documents.

This study examines the effects of vacuum freeze-drying on a wide variety of digital prints that were chosen because they broadly represent those often found in homes, offices or archival environments. In order to determine whether freeze-drying itself caused damage or whether changes that incurred in the prints were simply due to their preliminary wetting or subsequent freezing, the simpler methods of air-drying and drying in a blotter stack were also included in this study for the purpose of comparison.

2. Materials and Experimental

The samples consisted of prints made with nine different digital printing processes; within some of the processes, prints showing variations of inks and papers were selected. The 35 samples included inkjet (dye- and pigment-based inks on various coated and uncoated media), direct thermal (D1T1, monochrome and Polaroid ZINK), direct thermal transfer (D1T2), dye diffusion thermal transfer (D2T2), electrophotography (dry toner prints and liquid toner HP Indigo print), photothermographic transfer (Fuji Pictrography), thermal autochrome (Fuji Printpix) and digital exposure to photographic paper (gelatine silver and chromogenic). The processes and print materials have been described in detail elsewhere [2]. The prints were not made specifically for this experiment; instead they were taken from a larger collection of prints collected over the past ten years. They thus reflect a variety of dates and applications such as one might find in the holdings of a typical archive. Examples include printed office documents and scientific graphs, faxes, medical imaging prints, photocopies, reproductions and snapshots or documentary photographs. The samples were treated in four experiments that reflect common practice in water damage salvage efforts:

1. immersion and air drying
2. immersion and drying in a blotter stack
   2a. directly between blotters
   2b. between blotters but with a sheet of Bondina on the image side
3. immersion, freezing and air drying
4. immersion, freezing and freeze-drying in a vacuum chamber

In all procedures, samples were immersed in a bath of 18°C tap water for 10 minutes without movement, then removed from the water, held vertically and allowed to drip briefly. Care was taken not to touch the wet print surfaces. In experiment 1, samples were air-dried face up on blotters without first blotting off the surface. While air-drying avoids any potential interaction with the printed surface, which may be sensitive to the touch and/or tacky when wet, it has the disadvantage of taking up large amounts of table space – a factor that is critical when many objects need to be dried quickly in a restricted space. In addition, prints may cockle and curl while drying.

In experiment 2, samples were dried in a stack of blotters that was weighted with moderate weight, a method that allows for the drying of more samples on a smaller table space with the added advantage of resulting in flattened prints. In 2a, the wet print was in direct contact with the blotters on both recto and verso, but in 2b an interleaving sheet of Bondina (a thin, smooth, heat-bonded, non-woven 100% polyester fabric similar to Hollytex) was placed on the print surface to protect it from potentially becoming stuck to the blotter.

For experiments 3 and 4, the wet samples were placed individually on a rack, not stacked, and shock-frozen to -19°C in a freezer. In experiment 3, samples were then allowed to thaw and air-dry face up on blotters. In experiment 4, the frozen samples were freeze-dried in a vacuum chamber with a pressure of 1.6-1.8 mbar and without incremental heating.

The samples were examined visually before and after testing with the naked eye and under 10-40x magnification with the use of normal, raking, and specular light. Some samples were examined with ultraviolet (UV) radiation as well. Comparisons with untreated control strips were made. Evaluation of the results was qualitative, since no instrumental analysis was carried out.

3. Results and Discussion

The wide variety of processes and materials in this study produced results that vary greatly in type and extent. Within each process and each combination of materials (the latter being mainly relevant for inkjet printing, in which different inks can be used on different media), however, similar results were observed, so they have been summarised in table 1 and are discussed in the following.

3.1. Inkjet prints

The 15 inkjet prints reacted very differently to the experiments. Bleeding of the dye-based inks upon immersion into water was expected, since the inks were aqueous. Dyes can be quite waterfast on matte and microporous coated media, however. The pigment-based inks were waterfast regardless of media. The difference in water sensitivity between swellable and microporous coatings was very obvious: while the former were seriously damaged by immersion alone, with deterioration ranging from partial separation of two different layers to complete dissolution, the microporous coatings in general and that on a plastic film in particular withstood immersion and subsequent drying with hardly any physical changes. Following immersion, the matte and microporous coatings tended to yellow to varying degrees, the former more than the latter. This change was especially noticeable at the print edges. The cause of the yellowing is not clear – it may be the result of either the loss or migration of optical brightening agents, oxidative reactions of the coating's organic components, or both.

When examined under a microscope, the subsequent matte surface of the formerly glossy prints with swellable coatings that were frozen was found to be a result of the formation of minute craters in the coating; these may have been caused by the formation of ice crystals in the waterlogged layers. Freezing also caused two of the three swellable coatings to become opaque and flaky, leading to a white veil over the image and partial delamination.

The prints made with solvent-based inks on poly(vinyl chloride) banner media were not sensitive to any of the experiments; this is due to the non-solubility of the colorants in water and the lack of an ink receptor coating on the PVC. While inkjet prints made with solid inks were not tested, it may be assumed that the results would be similar to those of the direct thermal transfer prints, since they are similar in materials (colourant in waxy binder) and structure (ink on surface of paper or plastic).

3.2. Electrophotographic prints

In areas of high density it was found that the thick toner layer of the colour laser prints cracked when the prints were immersed into the water bath. Toner contains as a binder a blend of thermoplastic resins and is therefore hydrophobic. The paper support to which the toner is fused, on the other hand, expands when it becomes wet as a result of the uptake of water; this expansion is greatest across the grain, or fibre direction, of the sheet. Since the toner does not expand in the water bath, it relieves the accumulating stress caused by the expanding paper by cracking parallel to the grain direction. The difference in tension between the paper and the toner layer may also be the reason for the substantial concave curling of the wet prints; this curl remained, if to a lesser degree, in the dried prints, regardless of drying method.

Table 1. Summarised results of the experiments. "N/C" indicates: no significant changes observed.

Process and materials	Immersion	1. Air-drying	2. Blotter-drying	3. Freezing and air-drying	4. Freezing and freeze-drying
Inkjet (dye ink on uncoated paper)	Profuse ink bleeding.	Moderate cockling and curl.	Bleeding into blotter.	Ice crystal marks in areas of bled ink.	Ice crystal marks in areas of bled ink.
Inkjet (pigment ink on uncoated paper)	N/C	Moderate cockling and curl.	N/C	N/C	N/C
Inkjet (2 dye and 1 pigment ink on matte coated papers)	Dye: minor ink bleeding. Pigment: N/C	Slight to serious yellowing of coating.	Slight to serious yellowing of coating.	Slight to serious yellowing of coating.	Slight to serious yellowing of coating.
Inkjet (4 dye and 1 pigment ink on microporous media: 2 papers, 1 fibre-based paper, 1 plastic, 2 RC-papers)	Dye: occasional minor ink bleeding. Pigment: N/C	Occasional moderate yellowing of coating. Fibre-based paper: moderate curl.	Occasional moderate yellowing of coating.	Occasional moderate yellowing of coating. Fibre-based paper: surface is matted.	Occasional moderate yellowing of coating. Fibre-based paper: surface is matted.
Inkjet (2 dye and 1 pigment ink on swellable RC-papers)	Dye: profuse ink bleeding. Pigment: N/C Coating separates into 2 layers and partially disintegrates.	Minor curl.	Surface sticks to blotters and is matted by Bondina. Verso can also stick to blotters.	Upper and/or lower coating layer is matted, can become opaque and show ice crystal marks.	Delamination of opaque upper layer possible.
Inkjet (solvent ink on PVC banner)	N/C	N/C	N/C	N/C	N/C
Electrophotography (3 dry toner colour laser prints)	Printed areas do not wet, and thick toner cracks parallel to fibre direction.	Cockling and curl can be serious.	Minor curl.	Minor curl.	Minor curl.
Electrophotography (liquid toner HP Indigo print)	Printed areas do not wet, and strong curl.	Minor curl.	N/C	Minor curl.	Minor curl.
Digital Exposure (3 chromogenic, 1 silver gelatine fibre-based print)	N/C	Fibre-based print: moderate curl. Chromogenic RC papers: minor curl.	Surface sticks to blotters and is matted by Bondina.	Fibre-based print: moderate curl. Chromogenic RC papers: minor curl.	Fibre-based print: moderate curl. Chromogenic RC papers: minor curl.
Direct thermal (1 fax, 1 ticket, 1 bank statement)	Strong curl possible.	Moderate cockling and curl and slight yellowing of coating.	Slight yellowing of coating.	Slight yellowing of coating	Slight yellowing of coating.
Direct thermal (Polaroid ZINK print)	Topcoat becomes opaque.	Topcoat clears, reticulation of topcoat where water pools on glossy surface.	Topcoat clears, surface is matted by both blotters and Bondina.	Topcoat clears, reticulation of topcoat where water pools on glossy surface.	Formation of opaque crystals in topcoat.
Direct thermal transfer (1 colour, 1 black-and-white print on paper)	Printed areas do not wet.	Moderate cockling and curl.	N/C	N/C	N/C
Dye diffusion thermal transfer (2 prints with, 2 prints without wax topcoat)	Print does not wet, but edges may swell where water enters into support.	Matte deposits where water pools on glossy surface.	N/C	N/C	N/C
Photothermographic transfer (Fuji Pictrography)	N/C	N/C	Surface sticks to blotter, not to Bondina.	Surface is matted.	Surface is matted.
Thermal autochrome (Fuji Printpix)	Topcoat becomes opaque.	Topcoat clears, surface is matted.	Topcoat clears, surface is matted and sticks to blotter, not to Bondina.	Opaque topcoat with ice crystal marks.	Opaque topcoat with ice crystal marks, local delamination of topcoat.

3.3. Digital exposure to photographic materials

While photographic materials are not necessarily considered to be digital prints, the fact that today they are almost always exposed by digital means and are often present in archives along with the other types of prints was sufficient to include them in this study. The samples tested withstood short-term immersion well. Due to the increased surface tack of the swollen gelatine emulsion, the prints did stick to the blotters in experiment 2. They could be removed easily, but some paper fibres remained adhered to the surface. While this problem was avoided with the use of Bondina as a non-sticking, water permeable layer, the wet emulsion took on the surface texture of the Bondina and became matte when dry. The samples that were not physically restrained during drying (experiments 1, 3, 4) tended to curl due to tension between the emulsion and the support.

3.4. Direct thermal

Two of the direct thermal prints that were on a very thin paper support curled strongly during immersion. The prints retained their curl to a lesser degree when air-dried. Blotter-drying, freezing and freeze-drying did not appear to have negative effect on the prints except for an overall slightly irregular surface. All of the prints showed slight yellowing, and examination under a UV light source revealed that immersion caused a great loss or at least migration of optical brightening agents.

Polaroid ZINK (stands for "Zero Ink") is a direct thermal process which uses a special media that has three thermally sensitive layers, one each for yellow, magenta and cyan dyes. It is used for small-format amateur snapshots and may be regarded as a digital sequel of the earlier instant prints that were based on photographic technology [3]. Although the protective topcoat became opaque when wet, it cleared again upon drying. It did form a reticulation pattern where water pooled on the surface during air-drying, however, so contact with water beyond the immersion period of ten minutes may be damaging. The prints seemed to withstand immersion, freezing and air-drying well, but the freeze-dried samples suffered from opaque crystalline enclosures in the protective topcoat.

3.5. Direct thermal transfer

The binder used to hold the colourants in this process is a blend of waxes and resins and is therefore hydrophobic. The prints were not damaged by immersion and only suffered from moderate cockling and curl when air-dried. The blotter-dried, frozen and freeze-dried prints dried flat and without apparent changes.

3.6. Dye diffusion thermal transfer

Also known as dye sublimation prints, these samples did not wet in the water bath. This is due to the hydrophobicity of the materials on both sides of the prints: the image layer is a blend of synthetic, organic polymers such as polyester and polycarbonate, which can be further coated with a protective layer of wax following printing, and the support is usually a complex laminate of layers of synthetic fibres and plastic films. Some media may have a paper core; this may be the explanation for the swelling of the edges in two of the four different samples during immersion. Apart from the swollen edges, which did not regain their original form during drying, no changes were observed in the prints except for some superficial matte deposits where water had pooled on the prints that were air-dried.

3.7. Photothermographic transfer

While the Fuji Pictrography sample showed no changes in the water bath or after air-drying, it did slightly adhere to the blotters in experiment 2a. This is likely a result of the fact that the receptor layer contains gelatine and therefore swells and becomes slightly tacky when wet. Freezing the wet samples resulted in a overall change of the surface from glossy to matte. The original surface gloss was not regained during both experiments 3 and 4.

3.8. Thermal autochrome

The Fuji Printpix samples suffered changes in all of the experiments. The prints reacted similarly to the ZINK prints in that the protective topcoat became opaque when wet but cleared again upon drying. However, the originally glossy surface air-dried matte and adhered to the blotters, indicating that the wet topcoat was tacky. Freezing left the topcoat opaque and freeze-drying then caused it to flake off, revealing the intact image below.

4. Conclusions

Although this study set out to determine whether freeze-drying could be an option for drying wet digital prints, it turned out that the key factors that determined whether a print would survive water damage were first its resistance to becoming wet and secondly its sensitivity to freezing, this being a preliminary step before freeze-drying. With the few exceptions of some swellable inkjet media, thermal autochrome and ZINK, freeze-drying did not cause any further damage than immersion and freezing had not already caused. Thermal autochrome and ZINK are relatively rare processes, and only one of the three examined swellable inkjet media suffered from delamination of the ink receptor layer when freeze-dried.

Freezing damaged more of the prints than had been expected. The frozen inkjet prints on fibre-based paper, a popular microporous media for aqueous inkjet that imitates traditional silver gelatine fibre-based photographic paper, suffered from matting. One of the swellable inkjet media turned opaque and showed marks left by ice crystals within the areas of ink bleeding, as did the simple dye-based ink on uncoated paper. Finally, the originally glossy photothermographic transfer print turned matte, and the topcoat on the thermal autochrome print became opaque, thereby obscuring the image.

For inkjet, results show that pigmented inks and microporous coatings have a clear advantage over dye-based inks and swellable coatings when wet. Dye-based inks generally bled into any material in contact with them, although on microporous coatings bleeding was only minimal. The overall best results were found with pigmented inkjet on uncoated paper and also with solvent-based inkjet on PVC banner, dye diffusion thermal transfer, direct thermal transfer and electrophotography, processes in which some or all of the involved materials are hydrophobic.

Although some prints suffered from image disruption by damaged coatings, the majority of the tested digital prints generally remained structurally undamaged in this study; it can therefore be concluded that most digital prints can be frozen and freeze-dried following water damage. Since they are simpler and therefore less likely to cause unexpected damage, however, the methods of air-drying and blotter-drying are preferred when possible. Individual recommendations for drying prints of different processes should be always be taken into consideration; these are given in table 2.

As a result of this study and information from earlier publications, the following general recommendations can be formulated for best practice for drying wet digital prints:

- Identify the processes as best possible before water damage occurs or at least before selecting a drying method once the prints have become wet. If the process is unknown, choose air-drying as the potentially least damaging method. Although most air-dried prints will show curl and/or cockling, they can usually be flattened later by a conservator.
- Any wet prints that feel slightly tacky to the touch should be air-dried rather than dried in a blotter stack. This is especially important for inkjet prints on swellable media.
- Physical contact with wet surfaces should be avoided if possible.
- Drying in a blotter stack will reduce curl, however prints with sensitive surfaces may stick to the rough and fibrous blotter surface. In these cases, Bondina or a similar material can reduce this blocking; however some print surfaces may then become matted.
- Dye-based inkjet prints should be air-dried individually to avoid bleeding into other materials.
- Inkjet prints on fibre-based paper and on swellable media, as well as thermal autochrome and photothermographic transfer prints should not be frozen.
- In addition, thermal autochrome and ZINK prints should not be freeze-dried.

Table 2. Recommendations for drying wet digital prints[a].

Process	Air-drying	Blotter-drying	Freezing and air-drying	Freezing and freeze-drying
Inkjet (dye ink on uncoated paper)	++	–	–	–
Inkjet (pigment inks, matte and microporous coatings)	+	++	o	o
Inkjet (swellable coatings)	++	–	–	–
Inkjet (solvent ink on PVC banner)	o	o	o	o
Electrophotography	+	++	o	o
Digital Exposure	++	–	o	o
Direct thermal (monochrome)	+	++	o	o
Direct thermal (Polaroid ZINK)	++, blot surface first	+	o	–
Direct thermal transfer	+	++	o	o
Dye diffusion thermal transfer	+, blot surface first	++	o	o
Photothermographic transfer	++	+, only with Bondina	–	–
Thermal autochrome	++	+, only with Bondina	–	–

[a] ++ : best method
\+ : second best method
o : possible method but no preference
– : method not recommended

Continued research in this area could include the systematic study of cross-sections of the prints in relation to these experiments; this would result in a better understanding of the problems found with the reactions of some of the supports and coatings to wetting and freezing. In addition, chemical analysis of some of the coatings, for example of that of photothermographic transfer prints, for which technical literature is scarce, would be useful. It would furthermore be interesting to determine whether the method of drying has a long term influence on the mechanical properties of the supports and coatings. The application of a brittleness test to the samples treated in this experiment would be one example [4]. Finally, it would be interesting to compare the results of this study, in which the samples were frozen and freeze-dried individually, to samples freeze-dried within stacks of wet documents, which is perhaps a more likely scenario for archival materials.

Appendix

Materials and equipment:

- Blotters: Type 040, Klug Conservation, Walter Klug GmbH & Co. KG, Badeweg 9, 87509 Immenstadt, Germany, www.klug-conservation.com
- Bondina: Type H3228, Conservation Resources International LLC, 5532 Port Royal Road, Springfield, Virginia 22151, USA, www.conservationresources.com
- Freezer: Rivacold, Cool Italia GmbH, Schmidener Weg 13, 70736 Fellbach, Germany
- Freeze-drying equipment (condensation device I-20 and vacuum chamber): Martin Christ Gefriertrocknungsanlagen GmbH, Postfach 1713, 37507 Osterode am Harz

References

[1] Adelstein P Z, Burge D M and Hanna J F 2006 Recovery of water-damaged digital and traditional prints *Proc. 3rd Int. Conf. on Preservation and Conservation Issues Related to Digital Printing and Digital Photography* (London: Institute of Physics) pp 64–67

Image Permanence Institute 2004 *A Consumer Guide for the Recovery of Water-Damaged Traditional and Digital Prints* (Rochester, NY: Image Permanence Institute)

Jackson C 1989 A short research project into the permanence of thermal fax papers *The Abbey Newsletter* **13(8)** 133–34, 136

Norville-Day H 1994 The conservation of faxes and colour photocopies, with special reference to David Hockney's home made prints *Proc. Conf. Modern Works – Modern Problems?* (London: The Institute of Paper Conservation) pp 66–72

[2] Jürgens M C 2009 *The Digital Print: Identification and Preservation* (Los Angeles, CA: Getty Conservation Institute)

[3] ZINK Imaging: http://www.zink.com

[4] Salesin E, Burge D M, Adelstein P Z and Reilly J 2009 Brittleness of digital reflection prints *Proc. Conf. NIP25 and Digital Fabrication* (Springfield, VA: Society for Imaging Science and Technology) pp 138–41

Journal of Physics: Conference Series **231** (2010) 012006 doi:10.1088/1742-6596/231/1/012006

Colour changes in prints during long-term dark storage of prints

Carinna Parraman
Senior Research Fellow, Centre for Fine Print Research,
University of the West of England, Bristol BS3 2JT, UK

E-mail: Carinna.Parraman@uwe.ac.uk

Abstract. The most significant impact on colour fading in prints is exposure to light and air. However what happens to coloured prints during long-term storage in boxes, drawers and on shelves? Measurements of samples, printed in July 2005, stored in a range of light and darkened storage conditions have shown some interesting initial results. As more emphasis is placed on the effects of light, the dark stability of inkjet prints is relatively overlooked when considering how to preserve or store coloured prints. This study and presentation builds on previous research [1] and has concentrated on the changes to colour during storage. With reference to ASTM F2035 - 00(2006) *Standard Practice for Measuring the Dark Stability of Ink Jet Prints*, the Standards outline points out that whilst natural aging is the most reliable method of assessing image stability, materials and inks any data that is produced quickly becomes redundant; therefore accelerated aging is more preferred. However, the fine art materials in this study are still very much in circulation. The leading fine art papers, and pigmented ink-sets used in these trials are still being used by artists. We can therefore demonstrate the characteristics of colour changes and the impact of ink on paper that utilises natural aging methods.

1. Introduction

With the emergence of digital imaging technologies in the 1980s, there followed a desire to print high quality colour images for artists. When Graham Nash and Mac Holbert, of Nash Editions USA, began printing using a Scitex Iris (3047) printer they realised that whilst the Iris technology produced beautiful, rich and dense colour, the dye-based inks were fugitive. They found that the early inks, if left in daylight for a few hours would fade. [2]

In Bristol in the late 1990s work with the Encad Novajet also confirmed these fears. Tests revealed that dye based inks were of such a fugitive composition that when printed on standard coated cheap paper stock, they could fade perceptibly if left in the open air and even in a dark room overnight. The problem of colours changing in the dark suggested other important contributory factors such as ozone or gas fading, which might be generated by air circulated from fans, or an open window. As more digital work began to be printed onto coated papers, the question of the compatibility between papers and inks became more urgent. This has been an ongoing problem, echoed by Henry Wilhelm's paper *How long will it last? An overview of the light-fading stability of inkjet prints and traditional color photographs.* [3]

The most significant impact on colour fading in prints is exposure to light and air. However what happens to coloured prints during long-term storage in boxes, drawers and on shelves? Artists working at the Centre for Fine Print Research are engaged in a range of digital printing and paper construction. This has included large-scale printing, and might be combined with other materials, such as stitching, screen inks, other printmaking processes, embossing and the surface of the paper might be dampened. These artworks might be stored in portfolios, in drawers, dry-mounted and unframed. The majority of prints are stored in boxes, but in some cases even in the dark as paper and images began to discolour, the stability of the prints began to be questioned.

2. Background and previous study 2005-2006 [1]

The original study (2005-2006) measured the changes to the whiteness of paper. Prints were stored in different conditions, which were monitored and measured eight times over ten months. Since 2005, the storage of these prints has continued in the same conditions and a new set of measurement data has been collected (2009-2010). In order to gain a thorough understanding of the long-term effect of the storage of prints: in drawers, boxes, open on a shelf or in plastic the emphasis here has moved from the measuring the whiteness of the paper to comparing the colour patches at the top and bottom of the piles. The papers containing OBAs showed an initial rapid drop off in brightness, with relatively little subsequent change occurring. At the end of the first study period (2006), colours and papers were compared and in some cases the average data gave negative results, suggesting that instead of fading, colours and papers were darkening. Hypotheses and questions arose when analysing the data, the assumption being at the time, there were anomalies in the paper surface, human error or problems with the equipment or calibration.

Other hypotheses concerning darkening of the paper samples [4] suggest that it was the paper that could be the problem. Papanagiotou described Japanese paper as indicating a darkening or yellowing, suggesting that the fibre content can contribute to colour difference. In previous accelerated lightfast pilot tests (1999) the surface of coated papers became yellowed, although this has not occurred during any real–time tests. Different manufacturing processes, such as the texture, sizing and construction of the paper may also contribute to colour changes. Yellowing of paper during storage is also considered by Mitsubishi paper mills [5]

Wilhelm research concurs also that ink is not necessarily the main problem, but the paper can affect the stability of an image. Their findings also suggest that humidity can cause changes to prints, such as colour darkening and an increased chroma instead of fading. Humidity in the UK is problematic and with an increase in erratic weather conditions and damper summers then more emphasis should be placed on how prints are stored [6].

Saunders gave examples [7] of some characteristics of fading and changing in fine art pigments as a result of the impact of atmospheric conditions. Organic dye-stuffs and lake pigments are highly fugitive. For example, Rose madder is known to fade within a few months if exposed to strong light, although darker tones are more permanent. In a fresco, lime destroys madder completely. There are less common occurrences of fading in inorganic pigments, but are more likely to darken, which is due to the existence of salts in a humid environment. For example, red lead can darken when exposed to humidity and light. It can also lighten in fresco paintings due to an interaction with moisture and atmospheric pollutants. Also a result of the existence of, and the capillary action as salts are drawn through walls, azurite - a deep blue copper mineral - can turn green.

The objective therefore was to continue with the testing of existing pigment and dye samples. We decided to continue with the samples with dye based inks as well as pigment, as this is still being used for many desk-jet photo printers. This is due to the brightness of the colours and being able to achieve the appearance of a full colour image using just four process colours. However for this paper we shall report on just the pigment samples.

2.1. Brief description of differences between pigments and dyes

Pigments are finely ground coloured particles that are suspended in a carrier liquid. In contrast to dyes, which colour by staining materials, pigment particles sit on top of a material surface. However, in inkjet printing the objective is to maintain the dye and pigment colourant on the paper surface by using

Journal of Physics: Conference Series **231** (2010) 012006 doi:10.1088/1742-6596/231/1/012006

a receptor layer [8]. It is generally considered that permanent inks are usually made with pigments and non-permanent inks are made with dyes.[1] Furthermore, dyes are composed of organic materials that are more susceptible to fading than pigments, which are made from inorganic materials. Synthetic organic pigments tend to be more lightfast than their natural organic counterparts, they are highly saturated and can be used as an extender to the replace larger proportions of inorganic pigments; thus reducing manufacturing costs. Based on past experience of inkjet dyes of the past, early inkjet inks were dye based inks and faded in a relatively short space of time. However, as greater demands and with technological developments in high performance dyes and pigments, these definitions are beginning to be blurred. In very broad terms, many materials exist with a range of solubility and permanence; pigments at the low end and dyes at the higher end. Many definitions are not precise because of highly sophisticated manufacturing methods, for example pigments that have been ground to such ultrafine particulates could be considered as a stain. Furthermore, materials must be able to be resistant to light, moisture, air, alkalis, acids and also resistant to heat and pressure of an inkjet printer head system. These highly specialised materials are required for high performance applications that are very different to the traditional watercolours and paints as used by artists.

Some typical examples of organic and inorganic pigments are listed here: [9]

Organic: a) vegetable source (indigo, madder); b) animal source (sepia, cochineal); c) synthetic organic (phthalocyanine blue, alizarin crimson).

Inorganic: a) Native earth (raw umber, raw sienna); b) Calcined native earth (burnt umber, burnt sienna); c) Synthetic inorganic colour (titanium dioxide, cerulean blue).

There is a huge range of pigments from which to choose and for the inkjet chemist the important factor is to choose a colorant that when intermixed, the purity of colour is maintained. A cyan, magenta, yellow and black inkset may comprise a cyan ink that contains the chemical properties of a phthalocyanine pigment, the magenta ink composition may contain the quinacridone pigment, and the black ink is a carbon black, which is the most most resistant to fading and which can withstand visible and ultraviolet light, moisture, air[2].

2.2. Improvements in inkjet inks

Since 2000, inkjet inks have been developed for a higher degree of permanence. The cheaper and less permanent inks fulfil a necessary role for major sectors of the industrial and office market, and the high end fine art market inks have significantly shifted from dye based to pigment based inks. Pigment inks have brought the longevity of printed output in line with standards established in the high-end museum quality print market.

Although dye based inks present a more saturated colour gamut, it has been necessary from our bias at the CFPR towards high quality fine art printing to utilise the highest quality pigment based inks. Although pigment inks are less intense than dyes, they are more resistant to UV and gas fading. The pigment molecules are more complex and break down at a slower rate than the simpler, small molecule dyes. Whilst early pigment inks lacked intensity, advances have been made to improve the quality of these inks. This shift has developed in response to user demand for archiving and colour longevity, especially for the production of large print works, including exhibition, display and the fine art and poster market, where it is essential for colours not to fade.

For the medium and wide-format market, the initial four CMYK ink sets have recently been expanded to eight and twelve colour sets. For example Canon's LUCIA Pigment Ink Technology

---

[1] Some inkjet manufactures describe their inks-sets as pigmented although they add a proportion of dye to enhance the saturation of the colour.

[2] For more information on the main families of pigments go to:http://www.specialchem4coatings.com/tc/color-handbook/index.aspx?id=families also for longevity of artists pigments go to Bruce MacEvoy's Handprint website (2006) http://handprint.com/HP/WCL/waterb.html#PB27

contains twelve-colour inks (a mixture of pigment and dye colours) using CMYK and red, blue, green, grey, photo grey, photo cyan, photo magenta, and matte black. Hewlett Packard have introduced their Vivera range of colours to their new Design Jet Z series, which includes CMY plus light grey, grey, matte black and photo black, orange/red, blue, green, light magenta and gloss. By adding colours to the existing colour set to address gaps in the colour gamut, the inclusion of green has resulted in the light cyan being redundant.

Epson's UltraChrome colour system has included two different black ink modes - photo black and matte black; Canon differentiates between matte black, regular black, grey and photo grey. According to Canon, the combination of both grey and photo grey enables smoother transitions from light to dark. Hewlett Packard have included four grey/blacks, with the addition of a gloss. As part of their media profiling management system this can be switched on or off to enhance density and is an automatic component of gloss papers.

Epson UltraChrome ink incorporates a High-gloss Micro-crystal Encapsulation, in which, according to Epson literature, each pigment is coated in a resin to reduce the grouping of pigment particles. This is similar to Hewlett Packard's Vivera Electrosteric Encaplsulation Technology or EET; negative electrostatic charges are introduced within the resin layer, which coat the pigment particles and prevent them clumping together.

2.3. Developments in colour change testing

Over the last ten years, the quality of inks and papers has vastly improved, as has the print technology. Printer manufacturers, have guaranteed print longevity of more than 100 years. However, a test by Wilhelm established that the compatibility of paper and ink had a significant impact on the life expectancy of the image. For example, if the same manufacturer ink, paper and printer were used there was hardly any visible difference over 25 years. However, using a different paper with different inks and printer resulted initially in a reduced gamut image and image degeneration in less than 3 months. For artists using fine art papers on a range of machines, this could have important implications for the longevity of their prints.

An significant introduction to the standards for inkjet permanence is the testing for the dark stability of prints. In the introduction to the Standard, *ASTM F2035 - 00(2006) Standard Practice for Measuring the Dark Stability of Ink Jet Prints*,[3] suggests that natural aging is the most reliable method determining time and understanding the changes in stability. However it points out that materials and technology are quickly updated and replaced and therefore data can become redundant.

It seems however in the last few years the inkjet market has begun to mature and fewer dramatic changes have occurred. New fine art papers have been introduced and extra colours added to the ink-set, however for the artists' print market the majority of materials are still in circulation and therefore it is possible to undertake natural aging experiments. Our measurements undertaken in 2005 still have relevance.

In the standard *ASTM F2035 - 00(2006)* the suggestion is that Arrhenius testing is a good long-term predictor for long-term dark storage. This procedure involves increasing the temperature at a constant relative humidity. In order to test for long-term dark storage stability the Arrhenius testing method is undertaken by Wilhelm research, and is described in more detail here [6].

3. Methodology

Following the question *"What happens to the prints at the bottom of a pile?"* A method was devised to ascertain changes to the printed colour when displayed and stored in different conditions. Different types of papers (Somerset Enhanced, Somerset Satin (uncoated) Hahnemühle William Turner, Hewlett Packard Gloss) were subjected to a range of conditions (open on a shelf under low light conditions, wrapped in plastic, stored in an archival box). Piles of prints were arranged to simulate how a print

[3] www.astm.org/Standards/F2035.htm

might fare at the top of a pile compared to the bottom of a pile. These piles were placed in archival print boxes, open on a darkened shelf and wrapped in plastic print bags. Using a CMYKLcLm colour set, an edition of 10 prints per test pile were printed onto the four test papers. Each test sheet measures 100 x 150 mm and contains 91 colour patches: CMYKRGB from 5% -100% colour and 100% colour with 20% increments of black. A clear space is also included for measuring the spectral reflectance and changes to the whiteness of the paper.

Each paper was coded according to the ink, paper and placement. For example, a sample labelled: *ab301* describes a pigment inkset that is printed onto Hahnemühle William Turner paper, which is placed at the bottom of the pile.

Table 1. Codes used for position of the print and paper type

Position of pile	Paper
001 shelf top of pile	ad Somerset Enhanced pigment printer
002 shelf bottom of pile	bd Somerset Enhanced dye printer
101 plastic top of pile	aa Somerset Satin pigment printer
102 plastic bottom of pile	ba Somerset Satin dye printer
201 frame	ab Hahnemühle W Turner pigment printer
302 box bottom of pile	bb Hahnemühle W Turner dye printer
301 box top of pile	ac Hewlett Packard Gloss pigment printer
401 open air	bc Hewlett Packard gloss dye printer

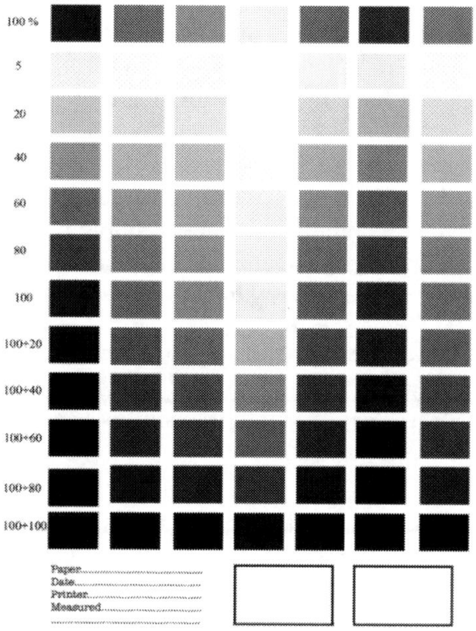

Figure 1. Test chart designed for CMYK printer.

3.1 Measuring (2005)

The original study began in July 2005 and using the Gretag Macbeth Profile Maker Pro measuring tool software, a custom target of 7 x 13 patches, including a space to measure the paper, was generated. From this the measuring data was saved as spectral and CIELa*b* formats. The first set of measurements was used as the reference data, and ten sets of measurements were undertaken over an

Journal of Physics: Conference Series **231** (2010) 012006 doi:10.1088/1742-6596/231/1/012006

eight-month period. Measurements were made to the paper and colour patches of the printed sheet on the top of the pile and the print at the bottom of the pile. Initial measurements for this study were taken in October 2009 and have been updated in February 2010. Over the course of measuring, initial negative values demonstrated early indications that colour and paper has darkened over time. Early suggestions for these negative results included human or measuring error, but as a result of re-measuring and rechecking data, and after 3 further years of storage the reflectances show a clear indication of darkening.

In the previous study and paper, the effect of optical brighteners was explored and considered against Tappi and CIE whitness. The whiteness differences between the papers was measured at the beginning of the test and at the end of the test period. At the same time, a series of measurements were undertaken to 100% colour patches along the top of the printed test charts.

3.2. Measuring (2009)

Although measurements were taken for the pigment and dye samples, for this paper, we will concentrate and report on pigment materials and printers. Therefore two fine-art inkjet receptive, one uncoated fine art printmaking and one photo gloss pigment receptive paper was measured. All of the prints are stored in the same room, which has both natural (non direct) light and fluorescent strip lighting. The room, and consequently the prints, have been subject to variable lighting and heating condition for 43 months. The temperature has therefore ranged from a hot humid summer day to a cold winder day with no heating. The Gretag Macbeth Spectrolino with UV filter standard illuminant of D50 and 2° colorimetric observer has continued to be used for these measurements. Differences are calculated between the reference data from July 2005 and the most recent set of data.

4. Results

As initial measurements and comparisons in 2005 showed a peak in the violet region of the photo-gloss paper sample, a UV filter was used for measuring all the paper samples. From the new data, we have been able to compare any changes occurring to colour patches at the top and bottom of the piles The following figures show the differences between the original (2005) measured paper sample (an unbroken line), a print at the top of the pile (identifiable by a line intersected by squares) and a print at the bottom of a pile (identifiable by a line intersected by circles). The results demonstrate that paper samples kept out of the light have not faded but a have darkened. In some cases this has occurred very slightly as demonstrated in samples kept in plastic. In other cases where samples have been kept open on the shelf samples have darkened considerably more. Samples open on top of the shelf had darkened more than the samples at the bottom of the pile.

5. Conclusion

The paper, through colour measurements and spectral graphs has compared the original data (2005-2006) and a new set of measurement data (2009-2010). All of the prints are stored in the same room, which has both natural (non direct) light and fluorescent strip lighting. The room, and consequently the prints, have been subject to variable lighting and heating condition for 43 months. The temperature has ranged from hot humid summer days to cold winder days with no heating. Whilst this does not present an ideal or quantifiable condition for the maintenance of prints, it does represent typical storage conditions in art and photographic studios, art departments, universities, libraries and homes. For archive conditions such as galleries, museums, archives and repositories it is unlikely that effects of ozone and humidity will impact collections.

The samples were kept in the same conditions as the previous tests: in drawers, boxes, open on a shelf and in plastic. As exampled, when comparing papers at the bottom and top of open piles, the colours and fine art papers (Somerset Enhanced, Hahnemühle William Turner, Somerset Satin) on the top of the pile had darkened. This might be due to the combination of ozone, humidity and dirt on the paper through the collection of dust. It is interesting to note that as shown in figure 2, which are details from figures 4 and 13, the Somerset Enhanced sample on the top of the pile darkened more than the corresponding sample on the bottom of the pile, whereas for the gloss sample the situation is reversed. This might be due to the lightening of the gloss paper sample. As shown in figure 3, which are details from figures 7 to 9, a comparison shows larger differences between Somerset Satin paper samples stored on an open shelf, than those stored in plastic or the box. The Satin (which is not coated) paper may have become dirty over time or the fibres in the paper evenly changed through aging. This also has a correlation to the colour changes for all the printed colours on the Satin samples, which appear to have become darker. As described by Saunders, there is a similar occurrence between contemporary printed samples and traditional dyes and pigments as used in paintings and frescoes; these react to salts and pollution in the air. Even for samples that are kept in boxes (albeit there are only minute changes), there is still enough humidity to cause an effect on both the paper and the coloured samples.

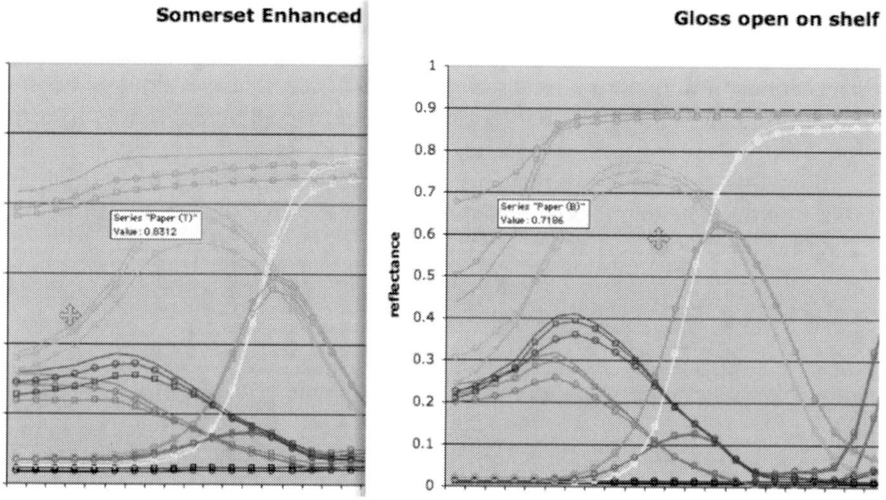

Figure 2. Somerset Enhanced papers on open shelf compared to gloss samples on shelf.

Figure 3. Comparison of Somerset Satin paper samples (top) open shelf, (middle) plastic, (bottom) box.

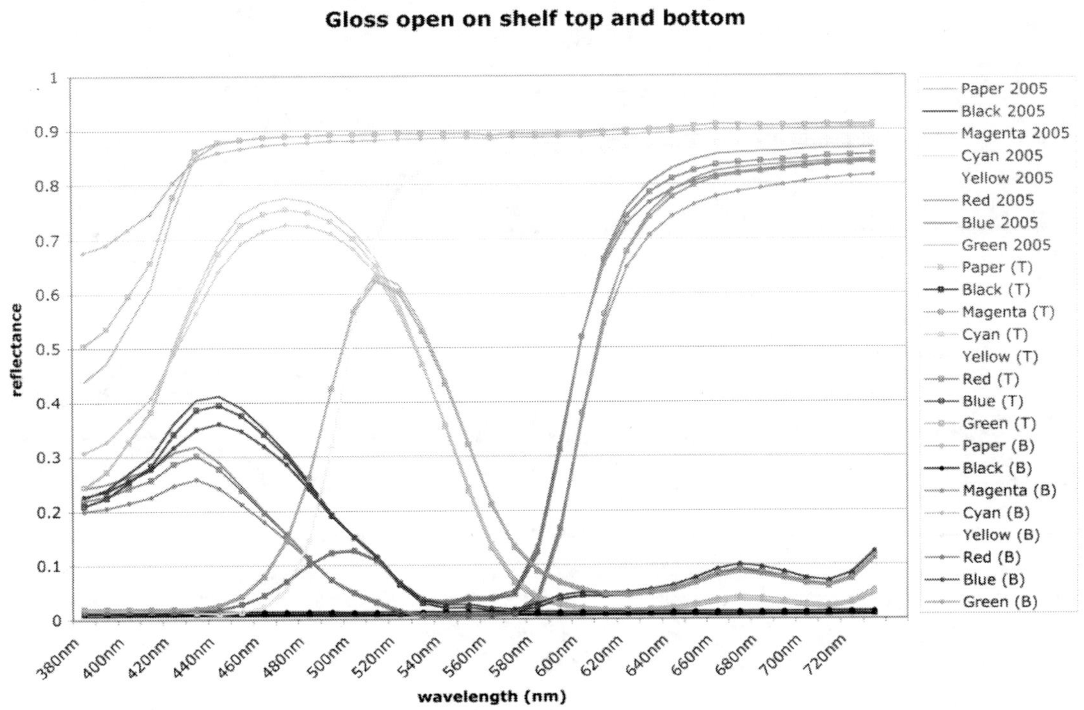

Figure 4. Colour measurements of gloss paper, open on shelf, top and bottom of the pile.

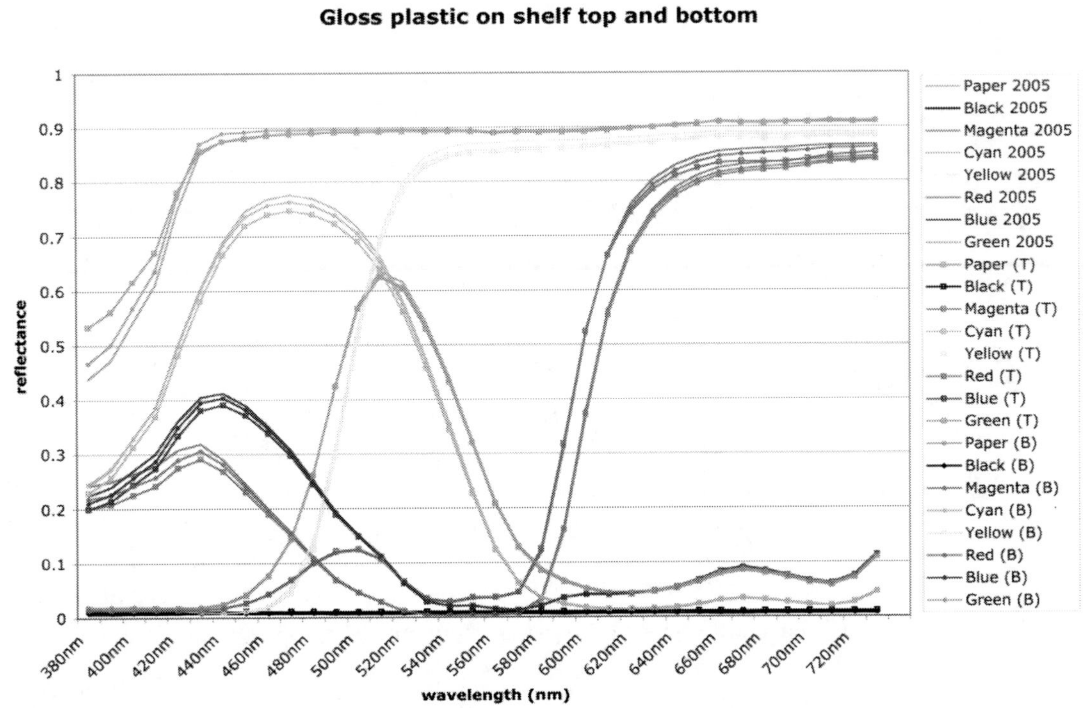

Figure 5. Colour measurements of gloss paper, wrapped in plastic on a shelf, top and bottom of the pile.

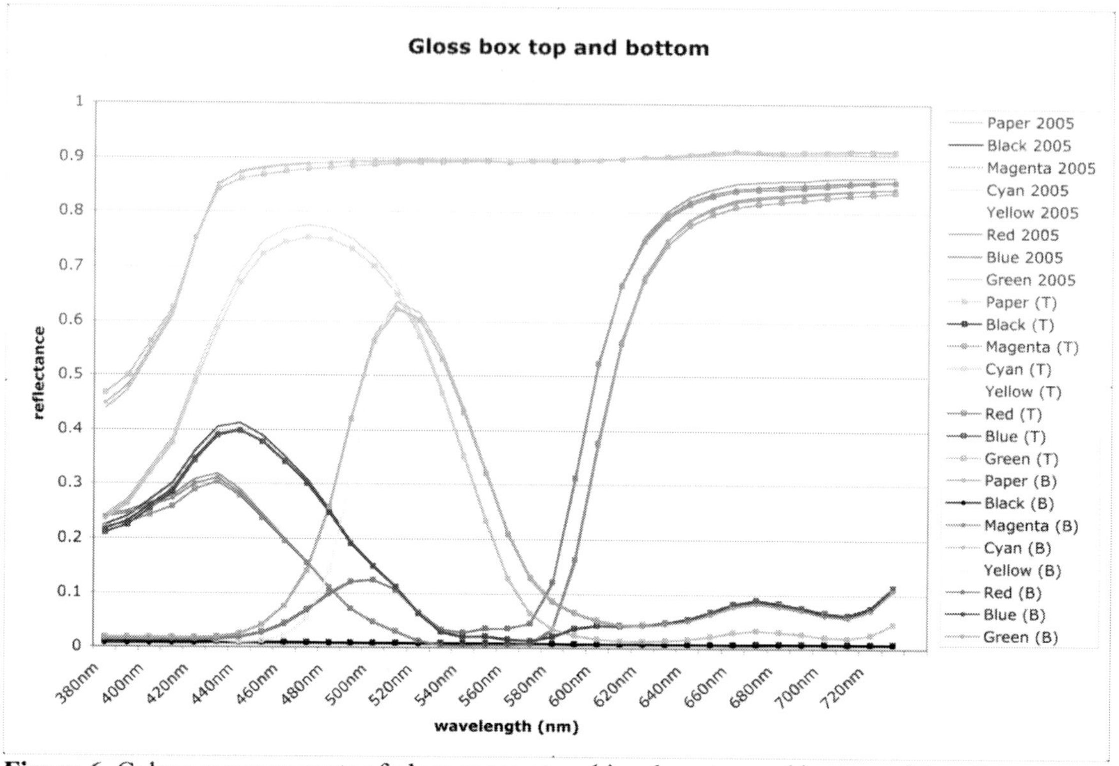

Figure 6. Colour measurements of gloss paper, stored in a box, top and bottom of the pile.

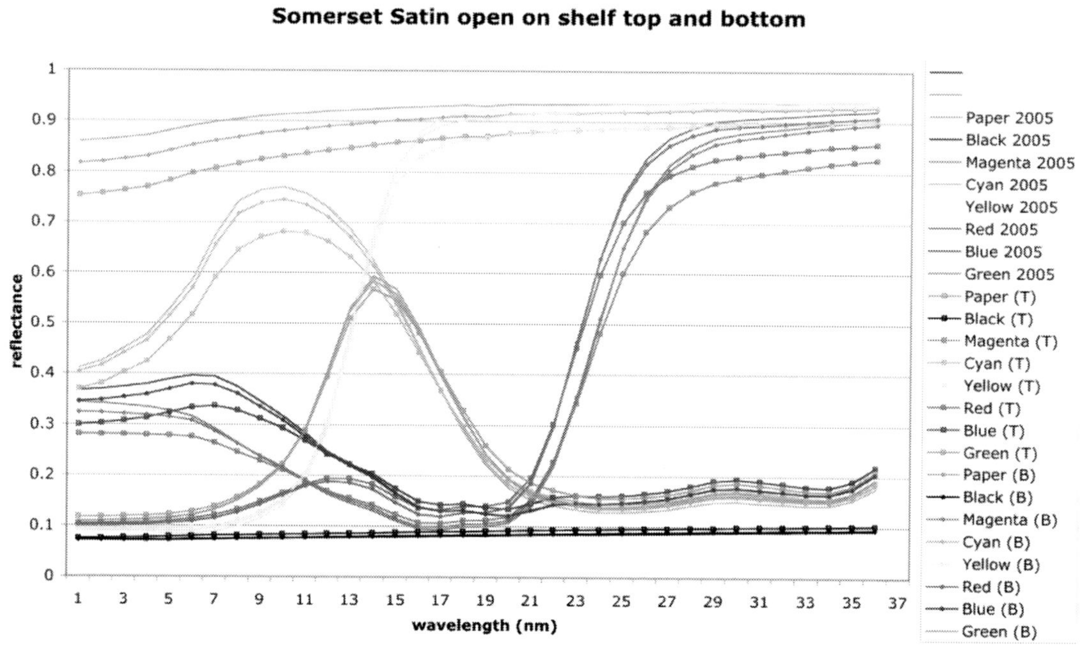

Figure 7. Colour measurements of Somerset Satin paper, open on shelf, top and bottom of the pile.

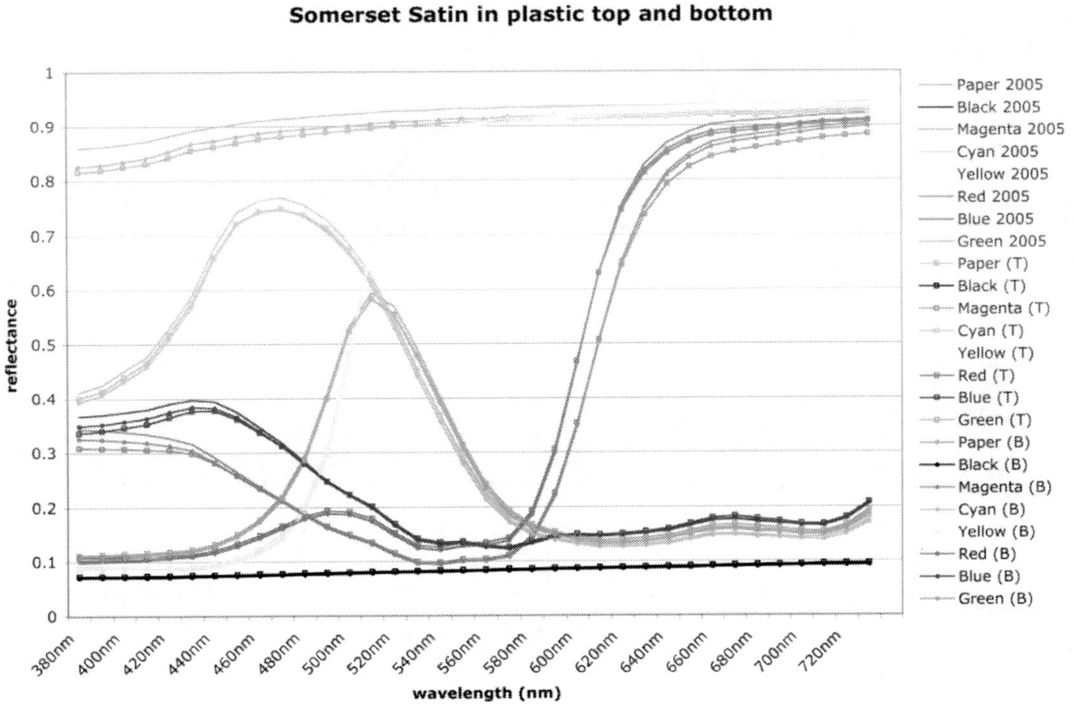

Figure 8. Colour measurements of Somerset Satin paper, wrapped in plastic on a shelf, top and bottom of the pile.

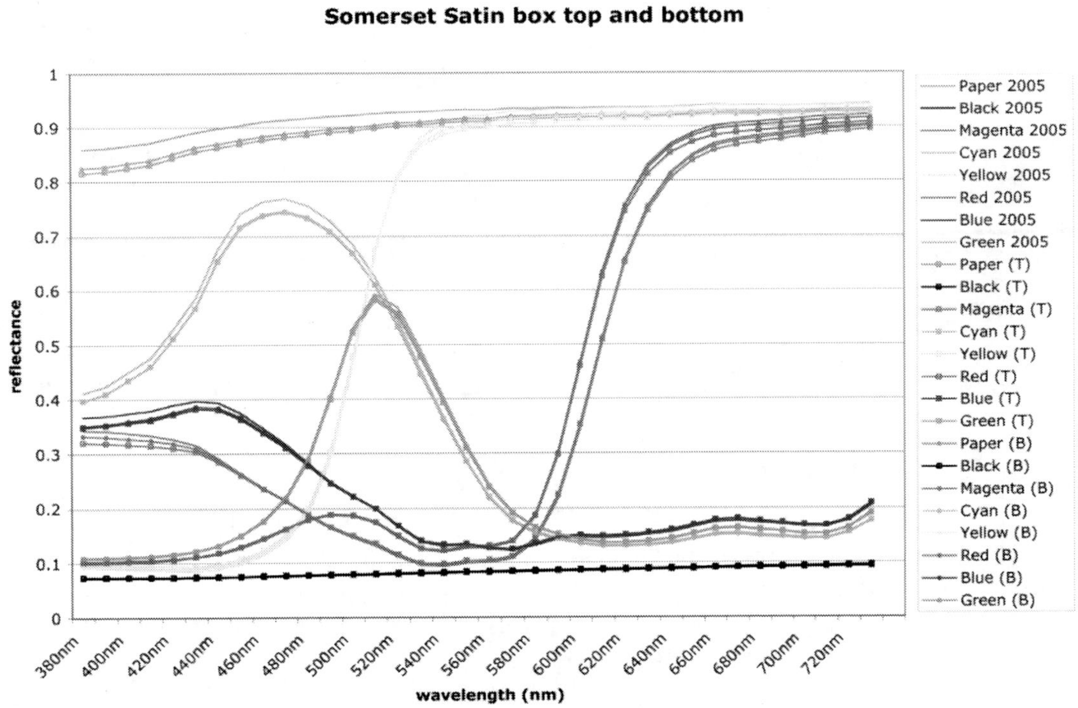

Figure 9. Colour measurements of Somerset Satin paper, stored in a box, top and bottom of the pile.

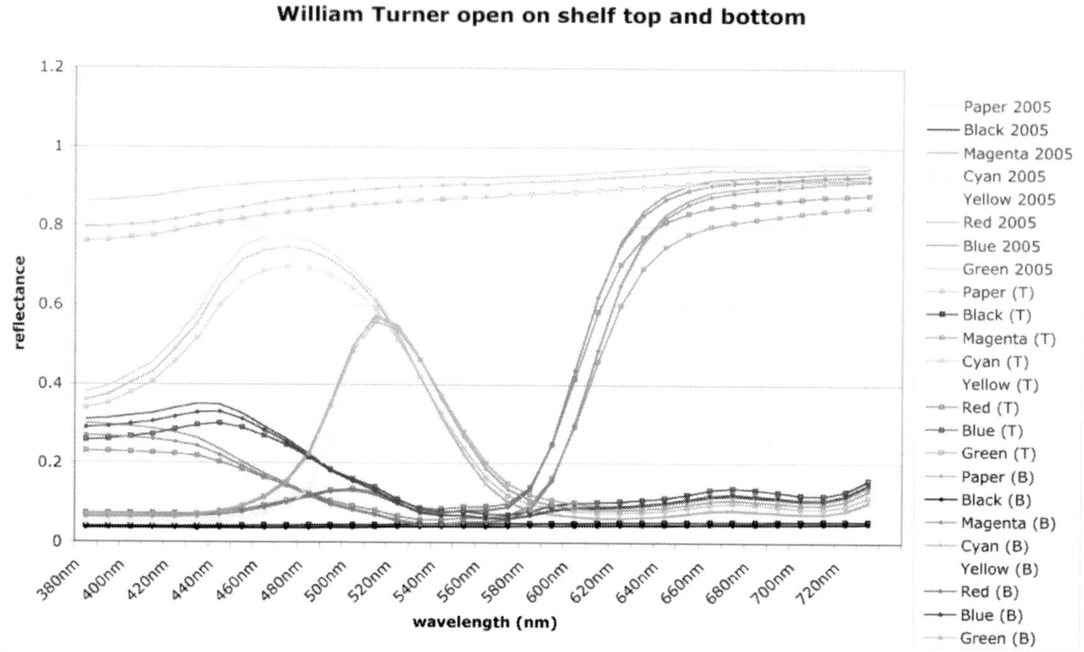

Figure 10. Colour measurements of Hahnemühle William Turner paper, open on shelf, top and bottom of the pile.

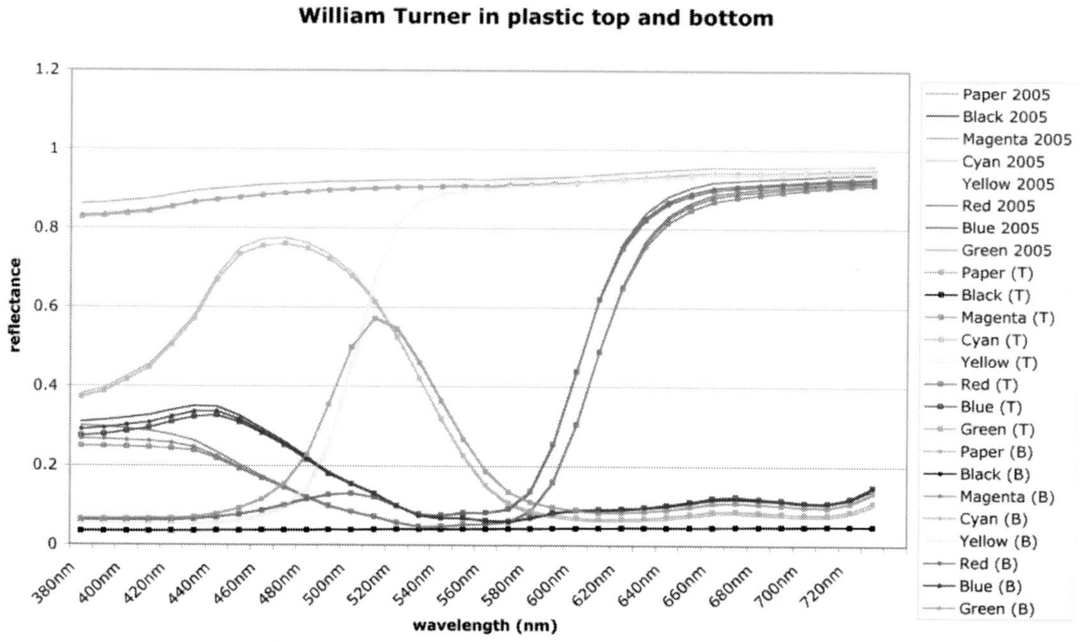

Figure 11. Colour measurements of Hahnemühle William Turner paper, wrapped in plastic on a shelf, top and bottom of the pile.

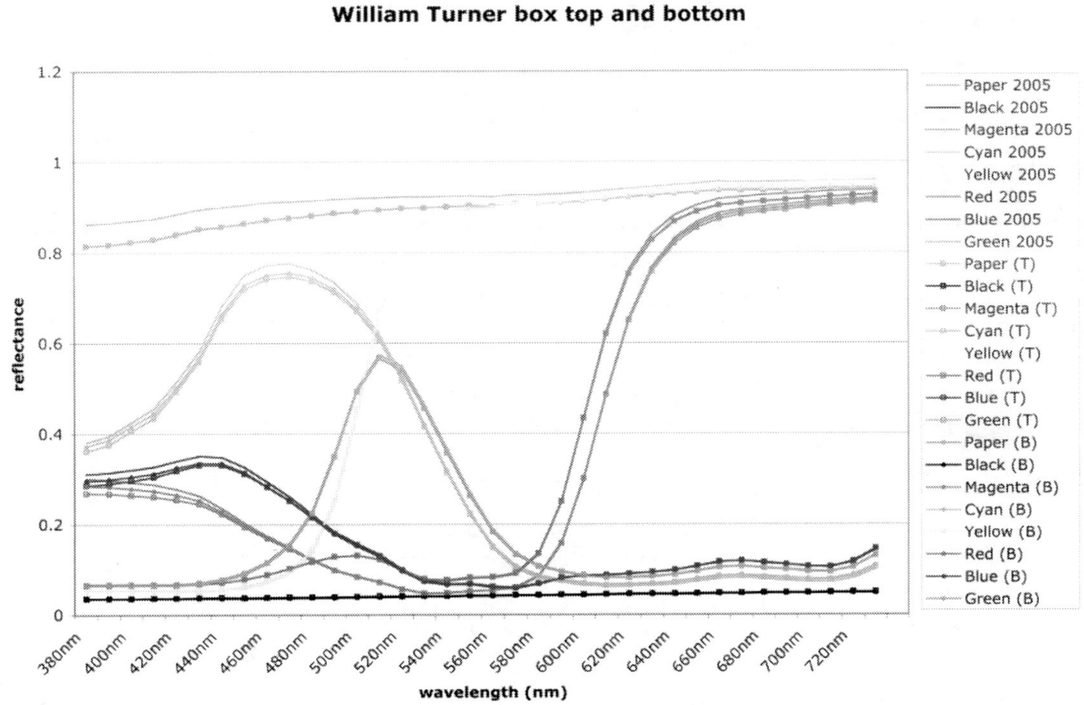

Figure 12. Colour measurements of Hahnemühle William Turner paper, stored in a box, top and bottom of the pile.

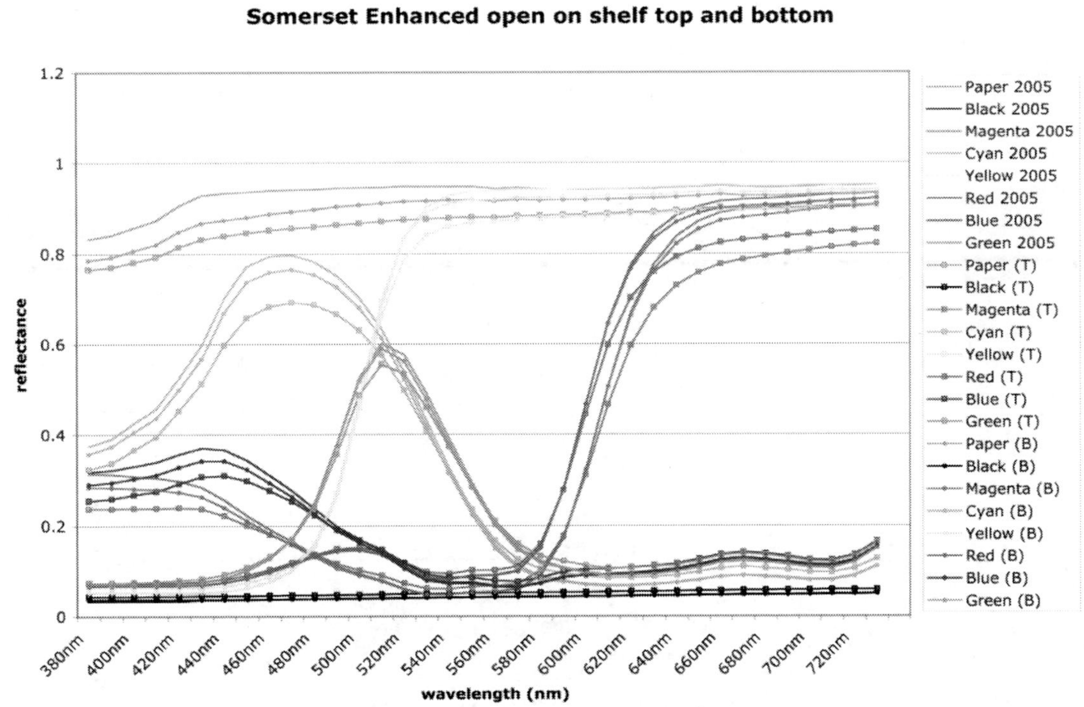

Figure 13. Colour measurements of Somerset Enhanced paper, open on shelf, top and bottom of the pile.

Journal of Physics: Conference Series **231** (2010) 012006 doi:10.1088/1742-6596/231/1/012006

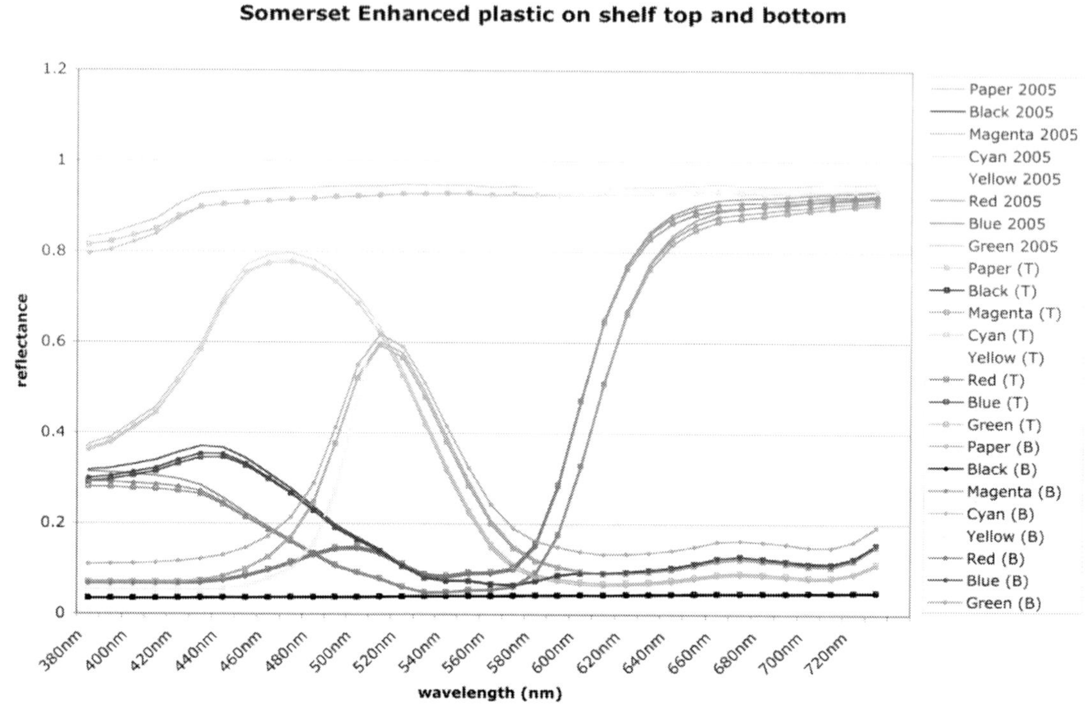

Figure 14. Colour measurements of Somerset Enhanced paper, wrapped in plastic on a shelf, top and bottom of the pile.

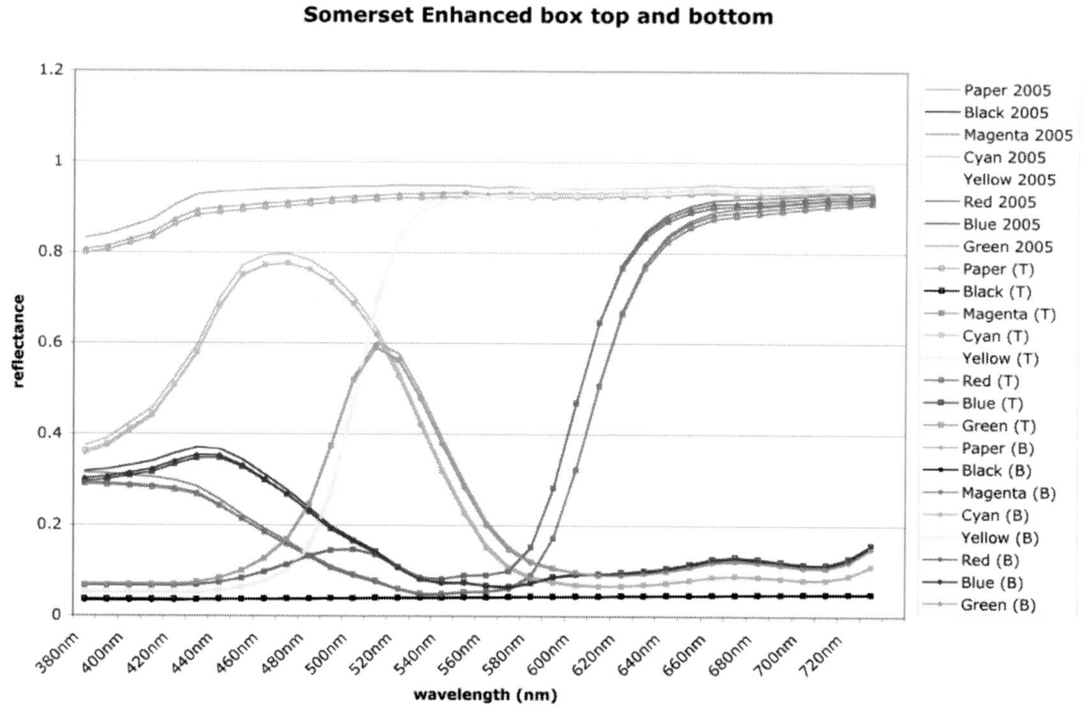

Figure 15. Colour measurements of Somerset Enhanced paper, stored in a box, top and bottom of the pile.

References

[1] Parraman, C. and Wang, H.Q. 2006. 'What happens to a prints during storage? To evaluate the whiteness artists' enhanced inkjet papers and changes in relation to the fading of ink' *Preservation and Conservation Issues Related to Digital Printing*, Institute of Physics, London, April 2006.

[2] White, G. (ed.) and Nash, G. 2006. *Nash Editions, photography and the art of digital printing*, New Riders Press.

[3] Wilhelm, H. 2002. 'How Long Will They Last? An Overview of the Light Fading Stability of Inkjet Prints and Traditional Color Photographs'. *12th International Symposium on Photofinishing Technology*, Orlando, Florida, USA, IS&T.

[4] Papanagiotou, C. 2000. 'An Investigation of the Use of Digital Colour Matching for the Toning of Papers using Ink-Jet Printing.' *The Paper Conservator.* Vol 20: 73.

[5] Mitsubishi http://www.mpm.co.jp/inkjet/english/storage.html

[6] http://www.wilhelm-research.com/ep9600%20print%20permanence.html

[7] Saunders, D. 2009. Colour changes in painted surfaces, *CREATE Colour Heritage and Conservation*, Università Degli Studi Di Milano, Italy, 19 – 24th October 2009, http://www.create.uwe.ac.uk/abstracts6.htm

[8] http://www.mpm.co.jp/inkjet/english/inkjet_media.html

[9] Hodges, E. 1989. *The Guild handbook of scientific illustration.* Guild of Natural Science Illustrators, Wiley. p.69-70

Selecting Suitable Enclosures for Digitally Printed Materials

D Burge[1] and L Rima

Image Permanence Institute at Rochester Institute of Technology, 70 Lomb Memorial Drive, Rochester, NY 14623, USA

E-mail: dmbpph@rit.edu

Abstract. It cannot be assumed that storage enclosures considered safe for traditionally printed images and documents are suitable for modern, digitally printed materials. In this project, a large variety of digital print types were tested using a modified version of the ISO 18916 *Imaging materials—Processed imaging materials—Photographic activity test for enclosure materials* standard to assess the risk to digital prints by paper enclosures known to be inert or reactive with traditional photographic prints. The types of enclosures tested included buffered and non-buffered cotton papers, and groundwood paper. In addition, qualitative filter paper that had been wetted and dried with either an acidic or basic solution was also tested to determine the effects of enclosure pH on digitally printed materials. It was determined that, in general, digital prints tended to be less reactive with various enclosure types than traditional prints. Digital prints were most sensitive to paper that contained groundwood. The enclosure reactivity test results were then integrated with previous published work on the tendencies of various enclosure types to abrade, ferrotype, or block to digital prints in order to create a comprehensive set of recommendations for digital print storage enclosures.

Introduction

The purpose of this project was to determine if various common paper enclosures will be harmful to digital prints, both documents and photographs, during long-term storage. To date, it has been assumed that what is safe for traditional prints will also be safe for digital prints. This is because digital prints are known to be sensitive to some of the same deterioration forces as traditionally printed materials namely oxidation of the colorants (dyes or pigments) and yellowing of the paper or plastic supports. No previous experimental work is known to have validated or invalidated this assumption. Digital prints may or may not be more sensitive to these decay forces than traditional prints, or they may be sensitive to as yet unidentified reactants within paper enclosures. This project rectifies this problem by providing data that addresses the issues.

Note that this paper assumes that inert plastic enclosure types such as polyester and polypropylene would be chemically safe for use in storing digitally printed materials. As such they were not included in the experimental work. However, it could not be assumed that there would be no harmful physical interactions between plastics and digital prints such as blocking, ferrotyping, or abrasion. These properties have been previously examined by IPI and published elsewhere [1, 2]. Those results will be integrated with the results from the present project to provide a set of recommendations for selecting housing materials that takes both the potential chemical and physical interactions into consideration. Certain plastics known to be sensitive to deterioration themselves and/or to emit harmful agents such

[1] To whom any correspondence should be addressed.

Journal of Physics: Conference Series **231** (2010) 012007 doi:10.1088/1742-6596/231/1/012007

as plasticizers or acids should never be used. These have already been excluded for use with photographs by ISO 18902 *Imaging materials – Processed Imaging materials – Albums, Framing, and Storage Materials* [3].

The results will be useful to anyone working in cultural heritage institutions charged with housing digitally printed collections materials. The data will also be useful to the ISO working group on the physical properties and permanence of imaging materials in the further development of their standards related to print storage.

Methods

The possibility of negative chemical interactions between traditional photographic prints and enclosure and display materials has been recognized for many years. For traditional photographic images, the Photographic Activity Test (PAT) was developed to screen for these potentially harmful enclosures. The method was subsequently standardized by the International Organization for Standardization (ISO) and is now designated as ISO 18916 *Imaging materials—Processed imaging materials—Photographic activity test for enclosure materials* [4]. This test involves the accelerated aging of special image interaction detectors and gelatin stain detectors against the enclosure materials being studied. The image interaction detector is a thin colloidal silver/gelatin layer on a polyester base that is used to screen for reactions that cause image fade. The gelatin stain detector is a photographic paper processed to minimum density (paper white) that is used to screen for reactions that cause print yellowing.

The current tests are based on that method but are modified from the original by the replacement of the image interaction and gelatin stain detectors with two new detectors made from the various digital printing materials. The image interaction detector is a uniform middle-grey (sRGB 128, 128, 128) print and the stain detector is unprinted paper (except dye sublimation which was printed to minimum density to include its overcoat and chromogenic silver-halide which was unexposed and processed to minimum density).

When printing, the "Best Photo" and "Photo Enhanced" printer settings were selected (when available) for photograph printing systems. Default settings were used for document printing systems. After printing, all samples were left to dry at 21°C and 50% RH in the dark for two weeks prior to testing.

The following enclosure materials were tested against the digital print samples:

- Non-buffered cotton paper
- Buffered cotton paper
- Groundwood paper
- Whatman No. 1 filter paper
- Whatman No. 1 filter paper soaked in 0.1N HCl
- Whatman No. 1 filter paper soaked in 0.1N NaOH

The hydrochloric acid (HCl) and sodium hydroxide (NaOH) samples consisted of Whatman No. 1 filter paper strips (the test control stipulated in ISO 18916) soaked in solution for one hour, and then left to air dry for 24 hours before testing. The papers were then tested to determine their pH using TAPPI T 509 *Hydrogen ion concentration (pH) of paper extracts (cold extraction method)* [5]. The HCl soaked filter paper had a pH of 3.9 and the NaOH soaked filter paper had a pH of 10.1.

Table 1 lists the printed digital materials that were used in the test as image interaction detectors. Multiple examples of each printer technology were tested when possible.

Table 1. Printed digital materials used as image interaction detectors.

Printer	Paper
Inkjet - Pigment – Photo 1	Fine Art 1
Inkjet - Pigment – Photo 2	Fine Art 2
Inkjet - Pigment – Photo 3	Fine Art 3
Inkjet - Pigment – Document 1	Plain Office
Inkjet - Pigment – Document 2	Plain Office
Inkjet - Pigment – Document 3	Plain Office
Inkjet – Dye – Photo 1	Porous Photo 1
Inkjet – Dye – Photo 2	Porous Photo 2
Inkjet – Dye – Photo 3	Polymer Photo 1
Inkjet – Dye – Photo 3	Polymer Photo 2
Inkjet – Dye – Document 1	Plain Office
Inkjet – Dye – Document 2	Plain Office
Inkjet – Dye – Document 3	Plain Office
Colour Electrophotographic 1	Plain Office
Colour Electrophotographic 2	Plain Office
Colour Electrophotographic 3	Plain Office
B&W Electrophotographic 1	Plain Office
B&W Electrophotographic 2	Plain Office
B&W Electrophotographic 3	Plain Office
Dye Sublimation 1	Dye Sublimation 1
Dye Sublimation 2	Dye Sublimation 2
Silver-halide – Colour	Chromogenic photo paper
Digital Press – Liquid Toner	Coated Glossy
Digital Press – Dry Toner 1	Coated Glossy
Digital Press – Dry Toner 2	Coated Glossy
Offset Lithography	Coated Glossy

Table 2 lists the unprinted digital papers used in the test as stain detectors. Note that many of the samples in table 1 used the same paper but were printed on different printers (i.e. the plain office paper). These papers were tested for stain only once. Again several examples of each paper type were tested when possible.

Table 2. Unprinted digital papers used in the test as stain detectors.

Paper
Inkjet Fine Art 1
Inkjet Fine Art 2
Inkjet Fine Art 3
Inkjet Photo – Porous 1
Inkjet Photo – Porous 2
Inkjet Photo – Polymer 1
Inkjet Photo – Polymer 2

Dye Sublimation 1
Dye Sublimation 2
Chromogenic photo paper
B&W photo paper
Plain Office
Digital Press - Coated Glossy
Offset Lithographic - Coated Glossy

The image interaction and stain detectors were cut into 2cm x 4cm strips in accordance with ISO 18916: 2007. All strips were read before and after incubation using a Gretag Spectrolino/Spectroscan (no UV filter, 2° observer, D50 illuminant) for CIELAB L*a*b*. The test was performed in duplicate. Each enclosure paper was incubated with four detectors, two image interaction and two stain. They were stacked into a specimen jig for incubation in the following configuration.

---------------- Weight to provide 500 Pa (including top glass)
---------------- Glass
---------------- Filter paper separator
---------------- Image interaction detector (face down)
---------------- Filter paper separator
---------------- Enclosure material
---------------- Uncoated polyester
---------------- Stain detector (face down)
---------------- Filter paper separator
---------------- Enclosure material
---------------- Glass

Filter paper separators between the enclosure material and the print samples are required by the ISO standard. This filter paper separator was not used in the untreated Whatman No. 1 filter papers tests. Another strip of filter paper was placed between the top image interaction detector and the glass to prevent sticking between the two.

Once the jigs were prepared, they were placed in an incubation chamber on racks with sufficient space in between to allow for air circulation. The tests (sample sandwiches in jigs) were incubated at 70°C and 86% RH for 15 days in an ESPEC ESL-3CA Humidity Cabinet. At the end of the incubation, the sample jigs were disassembled and the detectors were removed and remeasured. Delta E was then calculated.

It was understood that the high temperature and relative humidity test conditions required in ISO 18916 will likely cause colorant bleed for some of the digital print processes under investigation. However, it was believed that the inclusion of known, nonreactive materials (non-buffered cotton paper and Whatman No. 1 qualitative filter paper) in the experiments should still allow useful conclusions to be drawn. This project should be seen as a preliminary investigation.

Results

Table 3 shows the delta E values for the printed samples in contact with the cotton, cotton buffered, and groundwood papers.

Table 3. Delta E values for printed samples in contact with cotton, cotton buffered, and groundwood papers.

Printer	Paper	Cotton Non-buffered	Cotton Buffered	Ground-wood
Inkjet - Pigment – Photo 1	Fine Art 1	4	3	4
Inkjet - Pigment – Photo 2	Fine Art 2	0	0	1
Inkjet - Pigment – Photo 3	Fine Art 3	1	1	2
Inkjet - Pigment – Document 1	Plain Office	1	1	2
Inkjet - Pigment – Document 2	Plain Office	1	0	2
Inkjet - Pigment – Document 3	Plain Office	1	1	2
Inkjet – Dye – Photo 1	Porous Photo 1	2	1	2
Inkjet – Dye – Photo 2	Porous Photo 2	4	3	5
Inkjet – Dye – Photo 3	Polymer Photo 1	10	10	17
Inkjet – Dye – Photo 3	Polymer Photo 2	17	17	20
Inkjet – Dye – Document 1	Plain Office	3	3	4
Inkjet – Dye – Document 2	Plain Office	11	11	11
Inkjet – Dye – Document 3	Plain Office	9	9	10
Colour Electrophotographic 1	Plain Office	3	3	2
Colour Electrophotographic 2	Plain Office	2	2	3
Colour Electrophotographic 3	Plain Office	6	6	7
B&W Electrophotographic 1	Plain Office	1	1	5
B&W Electrophotographic 2	Plain Office	3	2	6
B&W Electrophotographic 3	Plain Office	3	4	4
Dye Sublimation 1	Dye Sublimation 1	2	2	1
Dye Sublimation 2	Dye Sublimation 2	4	4	4
Silver-halide – Colour	Chromogenic photo paper	1	1	1
Digital Press – Liquid Toner	Coated Glossy	1	1	2
Digital Press – Dry Toner 1	Coated Glossy	2	1	3
Digital Press – Dry Toner 2	Coated Glossy	4	4	4
Offset Lithography	Coated Glossy	3	3	4
	Total	*119*	*117*	*149*

Adding the delta E values for each enclosure paper type shows that, in general, there was little difference between buffered and non-buffered cotton papers with respect to image interaction of printed materials (see bottom row of table 3). The groundwood papers tended to be slightly more reactive which matches actual experience of traditional photos that have been stored in groundwood containing envelopes and boxes.

Table 4 shows the delta E values for the printed samples in contact with the acidic, basic, and untreated Whatman No. 1 filter papers.

Journal of Physics: Conference Series **231** (2010) 012007 doi:10.1088/1742-6596/231/1/012007

Table 4. Delta E values for printed samples in contact with acidic, basic, and untreated Whatman No. 1 filter papers.

Process	Paper	Untreated Paper	Acidic Paper	Basic Paper
Inkjet - Pigment – Photo 1	Fine Art 1	4	4	4
Inkjet - Pigment – Photo 2	Fine Art 2	1	0	1
Inkjet - Pigment – Photo 3	Fine Art 3	1	1	1
Inkjet - Pigment – Document 1	Plain Office	1	1	1
Inkjet - Pigment – Document 2	Plain Office	1	1	0
Inkjet - Pigment – Document 3	Plain Office	1	1	1
Inkjet – Dye – Photo 1	Porous Photo 1	2	2	1
Inkjet – Dye – Photo 2	Porous Photo 2	3	4	3
Inkjet – Dye – Photo 3	Polymer Photo 1	10	10	14
Inkjet – Dye – Photo 3	Polymer Photo 2	16	13	16
Inkjet – Dye – Document 1	Plain Office	3	3	3
Inkjet – Dye – Document 2	Plain Office	11	12	11
Inkjet – Dye – Document 3	Plain Office	8	9	8
Colour Electrophotographic 1	Plain Office	2	3	4
Colour Electrophotographic 2	Plain Office	1	3	2
Colour Electrophotographic 3	Plain Office	6	7	6
B&W Electrophotographic 1	Plain Office	1	2	1
B&W Electrophotographic 2	Plain Office	3	3	2
B&W Electrophotographic 3	Plain Office	3	4	3
Dye Sublimation 1	Dye Sublimation 1	1	1	2
Dye Sublimation 2	Dye Sublimation 2	3	4	4
Silver-halide – Colour	Chromogenic photo paper	2	3	2
Digital Press – Liquid Toner	Coated Glossy	1	2	1
Digital Press – Dry Toner 1	Coated Glossy	2	2	2
Digital Press – Dry Toner 2	Coated Glossy	4	5	5
Offset Lithography	Coated Glossy	4	3	3
	Total	*115*	*121*	*119*

The acidic and basic papers behaved very similar to the untreated Whatman filter paper. The pH of enclosures at these levels (3.9 and 10.1) seems to have had no effect on the stability of the image. This is not to say that acidic or basic enclosures should be allowable for storing prints, as the pH of paper has long been correlated with longevity. Therefore, while the image may remain intact, the support may still deteriorate over time. This experiment was performed only to determine the effects of enclosure pH on fade or yellowing and not other parameters such as coating integrity or paper durability. For that reason, paper enclosures should still meet the pH requirements of ISO 18902 of 7.0 to 9.5 ± 0.2.

Table 5 shows the delta E values for the unprinted papers in contact with the cotton, cotton buffered, and groundwood papers.

Journal of Physics: Conference Series 231 (2010) 012007 doi:10.1088/1742-6596/231/1/012007

Table 5. Delta E values for unprinted papers in contact with cotton, cotton buffered, and groundwood papers.

Paper	Cotton Non-buffered	Cotton Buffered	Ground-wood
Inkjet Fine Art 1	10	9	11
Inkjet Fine Art 2	1	1	5
Inkjet Fine Art 3	3	3	7
Inkjet Photo – Porous 1	2	2	4
Inkjet Photo – Porous 2	2	2	5
Inkjet Photo – Polymer 1	6	5	14
Inkjet Photo – Polymer 2	7	7	14
Dye Sublimation 1	1	1	1
Dye Sublimation 2	1	1	1
Chromogenic photo paper	6	5	12
B&W photo paper	11	10	21
Plain Office	1	1	5
Digital Press - Coated Glossy	6	5	7
Offset Lithographic - Coated Glossy	5	5	8
Total	*62*	*58*	*117*

Again the non-buffered and buffered cotton papers were similar to each other in terms of their potential for staining or yellowing digital printing papers. Most of the prints yellowed somewhat just as a result of the incubation, but the groundwood paper was considerably more reactive than the cotton papers and caused additional yellowing of most of the prints.

Table 6 shows the delta E values for the unprinted papers in contact with the acidic, basic, and untreated Whatman No. 1 filter paper.

Table 6. Delta E values for unprinted papers in contact with acidic, basic, and untreated Whatman No. 1 filter paper.

Paper	Untreated Paper	Acidic Paper	Basic Paper
Inkjet Fine Art 1	10	10	9
Inkjet Fine Art 2	1	1	0
Inkjet Fine Art 3	4	5	3
Inkjet Photo – Porous 1	2	2	2
Inkjet Photo – Porous 2	2	2	2
Inkjet Photo – Polymer 1	5	7	5
Inkjet Photo – Polymer 2	7	8	7
Dye Sublimation 1	1	1	1
Dye Sublimation 2	1	1	1
Chromogenic photo paper	5	6	6
B&W photo paper	11	13	10
Plain Office	1	2	1

Digital Press - Coated Glossy	5	6	5
Offset Lithographic - Coated Glossy	5	5	4
Total	*61*	*70*	*55*

As with image interaction, the pH of the enclosure paper had little effect on the yellowing of the print papers.

Conclusions

The following conclusions were reached regarding the potential for enclosure-induced image fading or paper yellowing:

- Digital prints are as sensitive to groundwood containing paper enclosures as traditional prints.
- The pH of paper enclosures has little effect on the fading or staining of digital prints.
- Digital prints should be housed in paper enclosures that meet the requirements outlined in ISO 18902 *Imaging materials – Processed Imaging materials – Albums, Framing, and Storage Materials.*

IPI has previously performed and published research on three other aspects of enclosure suitability for digital prints: ferrotyping, blocking, and abrasion [1]. While the above project assumed that plastics which meet ISO 18902 should be chemically safe for both traditional and digital photos, there is the possibility of physical interactions which may be harmful. Certain elevated temperature and humidity combinations can soften the gelatin coating on traditional photos resulting in conforming (ferrotyping) or bonding (blocking) of the photo to adjacent smooth surfaces such as plastic films or glass. Experimental work at IPI showed that all digital print types (inkjet, electrophotographic, and dye sublimation) were less likely to ferrotype or block than traditional photos. As such, plastic sheeting should be safe in terms of ferrotyping or blocking when used as an enclosure for digital prints, though recommended storage conditions (temperature and relative humidity) should always be maintained [6].

IPI has also performed research into the abrasion resistance of digital prints [2]. It was determined that abrading digital prints with common enclosures results in various types of damage including change to surface gloss, loss of colorant from the darker areas of the prints, and smearing of colorant into the light areas of the prints. While some print types were nearly impervious to abrasion such as color electrophotographic, others abraded very easily especially pigment inkjet prints on all types of papers. Some pigment inkjet prints may be so sensitive that no surface should ever be placed in contact with the face of the print. Polyester sheeting was the least abrasive enclosure material and is recommended for sensitive print types.

Given the above, it is recommended that all paper and plastic enclosures for digital prints meet the ISO 18902 standard. Additionally prints that are sensitive to abrasion should be housed in polyester sleeving as opposed to paper, and the most sensitive and valuable pigment inkjet prints should be housed in such a way that no surface comes in contact with the print's face. This can be achieved through the use of window mats or spacers when prints are stored stacked in boxes.

This project did not examine the potentially harmful effects of other materials used in housing digital photographs and documents, such as the adhesives used for constructing enclosures and writing instruments used for labeling. No recommendations on the use of these materials can be given at this time.

References

[1] Rima L and Burge D 2009 Tendency of Digitally Printed Materials to Ferrotype or Block *Technical Program & Proc. of IS&T NIP 25* (Louisville, KY, USA: Imaging Science & Technology) pp 142-145

[2] Nishimura D, Salesin G, Adelstein P and Burge D 2009 Abrasion of Digital Reflection Prints: The Abrasiveness of Common Surfaces and the Vulnerability of Print Processes *AIC Book*

and Paper Group Annual vol. 29 (Washington, DC, USA: Book & Paper Group)
[3] *ISO 18902 Imaging materials—Processed imaging materials—Albums, framing and storage materials* (Geneva, Switzerland: International Organization for Standadization)
[4] *ISO 18916 Imaging materials—Processed imaging materials—Photographic activity test for enclosure materials* (Geneva Switzerland: International Organization for Standadization)
[5] *TAPPI T 509 Hydrogen Ion Concentration (pH) of paper extracts (cold extraction method)* (Atlanta, GA, USA: Technical Asociation of the Pulp & Paper Industry)
[6] *ISO 18920 Imaging materials—Processed photographic reflection prints—Storage practices* (Geneva, Switzerland: International Organization for Standadization)

Acknowledgments
The authors would like to acknowledge the Institute of Museum and Library Services, who provided the funding for this project as part of a larger grant-supported research program to examine digital print preservation issues. We would also like to thank the Printing Applications Laboratory at the Rochester Institute of Technology for providing the digital press and offset print samples. In addition, Harman Technology Inc. and the Eastman Kodak Company provided printers and papers to support the project. The Image Permanence Institute is jointly sponsored by the Society for Imaging Science and Technology and the Rochester Institute of Technology.

Primary Author Biography
Daniel M. Burge, Senior Research Scientist, has been a full-time member of the Image Permanence Institute (IPI) staff for the last 19 years. He received his B.S. degree in Imaging and Photographic Technology from the Rochester Institute of Technology in 1991. He managed IPI's enclosure testing services from 1991 to 2004. In 2004, he took over responsibility for all of IPI's corporate-sponsored research projects. Currently he is investigating digital image stability and storage issues.

Journal of Physics: Conference Series **231** (2010) 012008 doi:10.1088/1742-6596/231/1/012008

Image Storage and Permanence Considerations in the Long-Term Preservation of Photographic Images – Update 2010

Joseph E. LaBarca

Eastman Kodak Company, 2400 Mt. Read Blvd., Rochester, New York, 14650-3024, USA

Abstract. Archivists and consumers, alike, need to become aware of long-term storage and preservation issues that relate to the preservation of the data behind digital photographic images. The more obvious issues, such as accidental or catastrophic data loss and hardware format evolution, are only now being recognized in the archiving community. Consumers need to be alerted to these issues and be prepared to develop preservation strategies as well. However, longer-term issues beyond routine backup and migration of data need to be considered. The very basic solution of preservation via hardcopy images stored in shoeboxes or albums is one option, but this raises a fundamental question regarding image preservation that transcends even the more complex solutions—the long-term stability of the chosen media, whether digital or analog. This paper discusses archiving and preservation as it relates to images, and the data behind those images, along with historical perspectives and an overview of possible longer-term preservation strategies [1–3]. The importance of image permanence standards, as they relate to overall selection of preservation strategies, will also be discussed.

1. Introduction

As digital imaging becomes more and more popular as the primary mode of capturing memories for the consumer, there is an ever-growing concern over long-term storage and the preservation of these memories. Despite growing efforts by the industry, the average consumer continues to be generally unaware that there is an underlying risk associated with image storage on computer hard drives or optical media. In many cases, because it is digital, the consumer actually feels that the images are safely preserved. Museums, conservators, and archivists, on the other hand, are starting to recognize the problem of digital image storage. Relative to preservation strategies, research and other published works are available on the topic. These strategies, however, are based on shorter-term storage with an associated longer-term migration plan. While some risk is mitigated, much remains.

2. Definitions

For the purposes of this paper, differentiation between image preservation and image archiving is needed. Archives are places where records and documents are stored. For digital image archiving, consider this to be a repetitive process of image file storage on the most current media and file format. As the formats change, a routine data migration plan is required to move the information to the next generation of media or file format. For digital image preservation, however, consider this to be a long-term, low-maintenance operation where the information is stored on a medium that is stable—

physically stable and stable in terms of format and readability. The need for frequent management of the data is highly reduced.

3. Historical perspectives

Many studies are available on the long-term storage of digital information [4]. The work has been driven largely by libraries, museums, and governmental institutions, and addresses the threats associated with computer-dependent systems. There is recognition of the need for ongoing data migration as hardware, software, and operating systems evolve. Because the quantity of information for these large institutions is huge, high-capacity systems are needed. A benefit of these systems is accessibility—the data is searchable and can be quickly retrieved. However, the cost of this accessibility is the dependence upon the computer and its associated storage devices, which continue to evolve, thus requiring a migration plan. An example of such a system is the on-line photo servers where storage is essentially free. Alternatives to these types of rapid access, computer-dependent systems have been studied where the ability to have rapid access is reduced in exchange for a reduction in the need for data migration. Examples have been discussed in the literature that make use of photographic film as the preservation medium [5].

Actually, the use of film for preservation is not new, as it has been done extensively in the motion picture and document industries for many years. Polyester-based, black and white silver separation films enabled the creation of separate red, green, and blue records of color motion picture that could last over 200 years in controlled, room temperature storage. The viability of this process has been demonstrated repeatedly in recent years with successful restorations of many classic films. Black and white microfilm provides stability up to and beyond 500 years at room temperature with good storage efficiency. Compression of documents in the order of 25 to 45 times is possible. New hybrid document imaging systems enable the use of film with scannable metadata for enablement of automated search and retrieval functions [6].

Preservation in the home has historically centered on hardcopy prints in albums, scrapbooks, and shoeboxes. There has been significant growth in digitally generated scrapbooks and photo albums in the last couple of years and growth of these products is expected to continue. Negatives, which are a somewhat compressed and more efficient storage format, are often times unorganized or simply not available. While a hardcopy print provides no compression of the information, and as such, tends to take up a large volume, it is human readable and therefore requires no system architecture to be put into use. Negatives and transparencies, on the other hand, do require some level of supporting system to use optimally, but they are human readable, nonetheless. They also provide for some compression of the data (4.2 times in 35 mm format, for example). Longevity of both print and film media will be discussed later in this paper.

As consumer imaging continues its rapid advance to the digital world, preservation remains erratic or non-existent. There has been little to no thought by the consumer for long-term preservation of their images. It seems they are taking for granted the automatic, "built-in" preservation that came with the traditional analog negative and print that was available for many decades. Today, digital files tend to be loosely organized on hard drives, CDs, and DVDs. But, often, there is no organization and little to no awareness of the vulnerability to loss of the image through hard drive crashes, media decay, and data corruption. Even if there is some level of organization, the consumer tends not to think about what will happen in the shorter term when they need to replace their computer, or the longer-term impact of file format and media format changes. All of these impacts result in the need for continual long-term migration of an ever-growing digital image collection.

4. Fundamentals of archiving

As mentioned above, archiving requires an ongoing commitment to manage records to keep them intact and ahead of any type of time-dependant changes. Key requirements of a good archiving system include low maintenance, cost-effective retrieval (based on the need for immediate or less-than-immediate access), as well as high storage density, longevity, and cost-effective life-cycle costs. What is an acceptable life cycle? The answer depends upon risk and balancing factors of hardware and media longevity. If one looks at 5.25- and 3.5-inch magnetic floppy disks, one could conclude a cycle of 20–25 years provides an acceptable risk, with note that the availability of computers with 3.5-inch drives has essentially disappeared over the last two years. However, looking at the more rapid evolution of optical media, one could conclude the time is well under 20 years. While consumers are now burning DVDs, there is a new, incompatible DVD format from the entertainment industry, with the Blue-ray DVD format having won out over the HD DVD format. An additional risk factor is the breadth of usage of a particular format, which can have both positive and negative consequences on the risk of obsolescence. CDs have been well established for many years by the music industry and could therefore be considered a long-surviving, low-risk format. However, unanticipated technology changes occur that can increase risk and shorten longevity. Music CDs are now under direct challenge from microdrives and flash memory in MP3 players. Will there be a full-scale format change? Consider VHS tape. It also became well established and actually rendered the Beta format obsolete. In 1990 it would have seemed likely that VHS was well established by virtue of the motion picture industry and would be low risk for a format change. Today, of course, VHS has become obsolete by DVD technology, driven by the same industry. So risk is always present and usually cannot be predicted far in advance of a format change.

Just as media format changes are hard to predict, technology advances around file structure and format will inevitably change as well. By its very nature, "technology" always advances. Improvements eventually render older technology obsolete. DVDs offered many significant improvements over VHS tape. Flash memory in MP3 players offers many advantages over mechanically driven CD players. Technology of file formats and structures continues to advance as well. With this in mind, why would we expect today's photo encoding format (JPEG) to endure? JPEG2000 already offers many improvements in compression over JPEG. The Windows Media Photo file format, later renamed HDPhoto, was announced in 2006 and shipped with the Windows VISTA operating system, but has not yet shown signs of replacing JPEG. With some fine tuning by the JPEG committee, HDPhoto has become JPEG XR and in mid 2009 became a published ISO standard (ISO/IEC 29199-2). With its improved compression algorithms, improvements in color reproduction accuracy, and support for High Dynamic Range (HDR) imaging, it is only a matter of time before camera manufacturers start to abandon JPEG. Will JPEG file formats be readable in 20 years time?

Being conservative on the risk from all sources, a 15–20 year migration cycle is very likely too long, and even half that time could be a problem. Depending on the media selection and how mature the system, using a five- to ten-year cycle is a better balance. A total rewrite of your database every five to ten years is a huge amount of work, however, especially when considering how quickly the volume of information grows. Total rewrite is a short-term consideration. One also must consider that consumers want the information to last for many generations, if not a lifetime or more, and this requires a very long-term commitment. Migration cycles are short because of the rapid and ongoing advancement of the computer and associated systems and the short life cycle of the media (see Figure 1.)

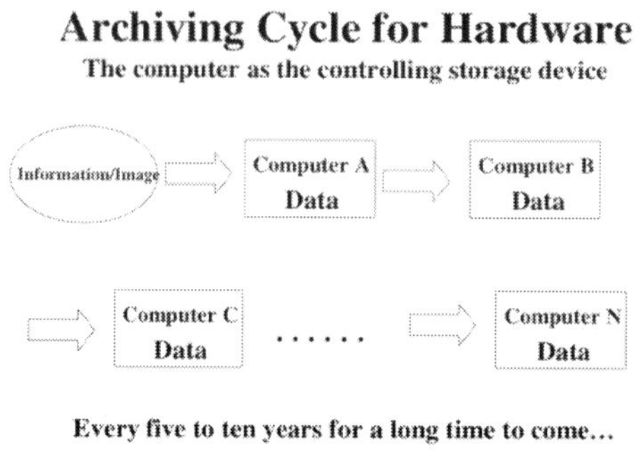

***Figure 1**. Information archiving with a computer-dependent system.*

In this system, data is migrated based on a cycle time driven by hardware and/or file format changes. Prior to any one of those components changing, a migration is required.

5. Fundamentals of preservation

Preservation takes a long-term perspective and attempts to reduce the need for short-cycle migration of the information. There are no simple solutions, however. The preservation system needs to be built to meet the user requirements, which is based on three broad-based options:

- Long-term need with immediate access: use an all-digital system and migrate frequently.
- Long-term need without the need for immediate access: use a hybrid-based system with an analog reference-archive backup.
- Long-term need that is low cost and simple to use: – use human-readable media with manual retrieval.

Reduction or elimination of the dependency on the computer can have a very positive impact on migration cycle time and allows for a preservation system that needs attention very infrequently. The key is to move from a digital storage format to a human-readable optical format using media that is very stable. Key requirements mentioned above (storage density, longevity, etc.) still apply with the addition of human readability. The computer is not eliminated, per se, but it becomes a gateway to create and retrieve objects containing the human-readable information, see Figure 2. A computer is actually not truly needed at either end of the cycle; although it can, along with the writer and scanner, enable greater operational and storage efficiency. In a case where a computer is unavailable, the document can still be retrieved using first principles: the document was in a human-readable format to begin with, and visual retrieval of the information can be accomplished using only a light and a lens.

Journal of Physics: Conference Series **231** (2010) 012008 doi:10.1088/1742-6596/231/1/012008

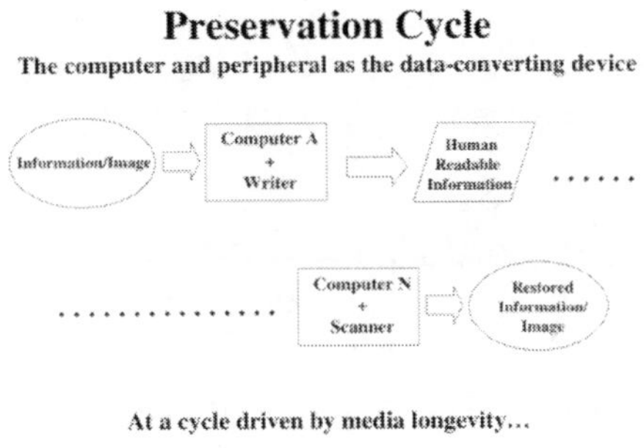

***Figure 2**. Information preservation with a computer-independent system.*

Peripherals to enable enhancement of the system include a film recorder and a film scanner. Both of these devices are well established and of high quality today. Film scanners have reached a state-of-the-art position and their evolution has slowed, while film recorders continue to grow in capabilities such as speed and resolution. This is being driven primarily by the motion picture industry. Film has also reached a stable, high-quality position for both image structure and resolution, as well as for long-term image stability. This makes it an ideal candidate as a storage medium.

6. Preservation by consumers today

While a film-based system has not yet been designed to work as an integrated system for high-volume storage of consumer digital files, options exist today for the consumers to store their images for the longer term. The first and simplest way is to make hardcopy prints of the most valuable images. Media of various technologies exist today to provide over 100 years of storage life at room temperature conditions. Printing is easily accomplished at home using photographic-quality 4 × 6 or page-sized printers, at retail locations using self-service kiosks or a retailer-operated digital minilab, or online. A second option available for storage is use of an online service such as the Kodak Gallery or HP's Snapfish. These services make use of digital storage devices to provide ready access to images; concerns about format migration are the responsibility of the service provider and should not be a concern of the end consumer. In most cases, this storage is available for free, if a minimum annual quantity of printing is done. A third, self-managed option is to use multiple magnetic hard drives with redundant backup. This option requires a disciplined approach and gets complicated as the collection of images grows in size. Making use of redundant drives is critical, recognizing the mechanical limitations and wear impact of rapidly rotating disk drives. A fourth, equally complex and less desirable storage option is to routinely move images to optical media such as CDs or DVDs. Three critical concerns in this option, as discussed above, are the media longevity, the hardware format longevity, and the file format longevity. The end consumer should set up and commit to maintain a long-term routine to rewrite the images to the latest media and format. A cycle of no longer than five years is recommended. While CD and DVD optical media is becoming available with advertised longevity of 100 years and higher, there is a growing amount of anecdotal evidence being gathered on low-cost CD media that lasts for two years or less. Any short-term plan by the end consumer to use optical media should center on higher quality, higher cost media only.

7. Stability of materials used for preservation

Table 1 outlines reflection and transmission media for use in preservation and provides information on their advantages and disadvantages. Reflection media has the advantage of being excellent for human readability. However, it is poor for storage density (albums fill up fast) and has variable longevity, which is technology dependent. Nonetheless, should the file ever become lost, a hardcopy

Journal of Physics: Conference Series **231** (2010) 012008 doi:10.1088/1742-6596/231/1/012008

print is a human readable record of the digital file. The industry needs to encourage people to make prints as preservation records of their images. Transmission media has variable unaided human readability, depending on image compression, which is a direct tradeoff to storage density. Depending on film type, the media has good to excellent storage longevity.

Table 1. Characteristics of reflection and transmission media for preservation.

Media Type	Human Readability	Storage Density	Longevity
Reflection	Excellent	Poor	Excellent to Poor
Transmission	Poor to Excellent	Excellent to Poor	Good to Excellent

Tables 2 and 3 discuss the longevity of various reflection and transmission technologies, respectively, and provide specific examples and longevity information for Kodak media [7, 8]. With one noted exception, the lifetime estimates in these tables use a new endpoint criteria that is more conservative than that typically used for consumer images [9, 10]. Because the application is for long-term storage, these predictions are based on dark keeping applications, and they include the effects of heat, humidity, and atmospheric pollutants; but they do not include effects of light.

Table 2. Stability of reflection media for long-term storage applications.[a]

Media Type	Media Name	Estimated Longevity	Comments
Thermal Dye Diffusion Transfer	*Kodak Professional Ektatherm XtraLife*™ media	80–100 years for 5% dye loss[b]	Virtually no sensitivity to humidity or ozone
Inkjet/Swellable Media	*Kodak Ultima* picture paper	Over 75 years	Using high-quality pigmented inks; media and inks continue to evolve
Inkjet/Porous Media	*Kodak Ultra* Premium picture paper	50 to over 75 years	Using high-quality pigmented inks; long-term testing in progress
Silver Halide	*Kodak Professional Endura* papers	Over 150 years	Virtually no sensitivity to humidity or ozone; greatest longevity of any silver halide paper

[a]Using a much tighter dye loss criteria of 15%; room temperature storage conditions of 23C and 50% RH and pollutant-free air; lower storage temperatures can provide significantly longer longevity
[b]Time for 15% dye loss has not been determined due to the extremely high thermal stability; actual time will be well beyond 100 years

Table 3. Stability of transmission media for long-term storage applications.[a]

Media Type	Media Name	Estimated Longevity	Comments
Silver halide color reversal film	*Kodak Ektachrome* films (E-Series)	Over 100 years	*Kodak Ektachrome* duplicating film has the highest image structure, lowest ISO speed of the *Ektachrome* family
Silver halide color reversal film	*Kodachrome* films	Over 120 years	Similar benefits to *Kodak Ektachrome* films; limited processing availability
Silver halide color negative film	*Kodak Vision* color intermediate film (motion picture)	Over 100 years	Image structure comparable to *Kodachrome* film; limited processing
Silver halide color negative film	*Kodak Professional Portra* films	Over 50 years	Image structure slightly reduced vs. *Kodachrome*, but high ISO speeds; processing widely available
Silver halide B&W negative film	*Kodak* black and white films	Over 500 years[b]	Silver image as opposed to dye; includes motion picture, micro-, and camera films

[a]Using much tighter dye loss criteria of 15%, room temperature storage conditions of 23C, 50% RH, and pollutant-free air, lower storage temperatures can provide significantly longer longevity
[b]When used with polyester-based films.

Evident in Table 3, there are significant longevity benefits for silver halide film. This makes sense because the design criteria for most films relative to image permanence has been to optimize for dark storage. In addition, film provides benefits in that it is a very mature technology, with few unknowns. Because film has been around for so long, there has been good confirmation of longevity predictions with real-time, natural aging data. In addition to its maturity, color films provide increased storage capacity because of their ability to record three separate channels of information, and they offer excellent image structure for high resolution and low noise. Color reversal film provides an additional benefit of its black surround (unexposed area), which reduces flare in viewing and scanning operations. Because they are gelatin based, these silver halide films all offer high resistance to humidity and atmospheric pollutants.

A long-term preservation solution was proposed and patented by Eastman Kodak Company in 1996 [11]. Using polyester-based microfilm, three color-separation files are created, and those files are written as a luminance channel and red and blue chroma channels, onto the microfilm within the space of a traditional 35 mm frame. Additional information within the frame is a calibration gray scale and human- and machine-readable metadata, see Figure 3. Additional details on this method were presented at the Imaging Science and Technology's Archiving Conference in San Antonio, Texas, in April 2004 [12]. Further information on digital file color encoding specifications and transformations was presented at the 2005 Archiving Conference in Washington, DC, on April 2005 [13].

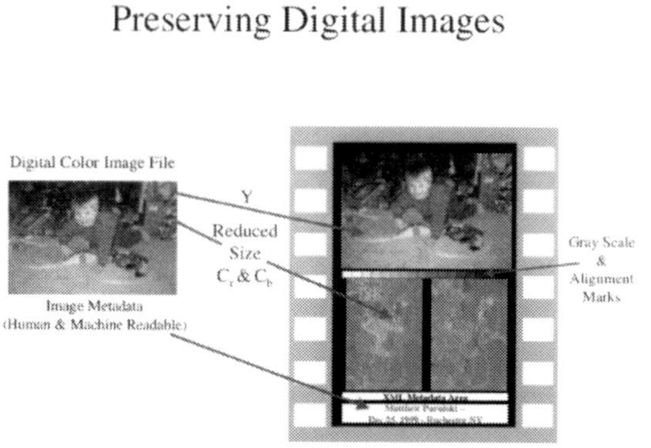

Figure 3. *Patented method for long-term preservation using black and white microfilm.*

8. The importance of metadata preservation

Metadata is, quite simply, data about data. Virtually all digital cameras today record a wide range of metadata when an image is captured, covering simple information like the date and time the image was taken, to information about the specific camera and actual camera settings used to capture the image. In cameras with GPS sensors, the location of the image can be included in the metadata information package as well. For professional and commercial photographers, efforts by the International Press Telecommunications Council (IPTC) established specific categories or "headers" for information to be recorded about the photograph beyond what was recorded by the camera. In 2001, Adobe Systems Incorporated created a new metadata framework called Extensible Metadata Platform (XMP) to allow cross-synchronization with the IPTC headers and newer IPTC core framework [14]. While professional and commercial applications of metadata are beyond the scope of this paper, it is important to understand that metadata can be edited, added to, or lost from the digital file of a photograph.

From the perspective of long-term storage and preservation of consumer digital images it is very important to understand the value of metadata and the risks of losing the metadata associated with long-term preservation. From the value side, metadata replaces the human "recorder" of information—the information previously written on the back of photographs, such as date and time the picture was taken. In the digital and computer world, metadata allows for rapid sorting of information by a multitude of categories. Examples of the more commonly used categories include file size and type, date modified, date picture was taken, author, title, and camera model. Less common attributes include picture dimensions, owner, and copyright information. On higher end digital cameras, specific information such as exposure and ISO settings, lens focal length, exposure and flash compensation, color space, and white balance are available as well. As mentioned above, if enabled, GPS location information is also available. All of this information is available for sorting and is often used in photo-album software for picture grouping and sorting. From the risk side, users need to be aware that this data can be corrupted or lost when making successive copies of the digital images during routine migration. The data can be lost as image files are moved in and out of certain photo editing and albuming programs (a short-term concern) or when file formats change (a concern primarily for the long term). For example, Exchangeable Image File Format (Exif), which is a specification for image file format including how and where metadata is stored, is supported by JPEG and TIFF but not by JPEG2000 file formats. Users of these types of software and file formats need to be certain that metadata remains intact.

Because of the value it brings to digital photography, metadata needs to be considered for long-term preservation outside the digital domain. In the film-based preservation scheme mentioned in the previous section, writing of metadata is taken into account in both human-readable and machine-readable formats. While it may not be needed for a consumer-oriented preservation scheme, recording all of the metadata contained in the original digital file may require a dedicated frame adjacent to the human-readable image. In the hardcopy media-based preservation scheme, recording the metadata is more problematic. Key information can be added into the image itself (shown as actual numbers such as some cameras do with a date function); doing this on all files in a collection would certainly be time consuming. Manually writing the information on the back of the print is the other alternative, which is also time consuming. So in addition to the low storage density offered by reflection media, preservation of metadata also needs to be recognized as a disadvantage, although not an insurmountable one. For those most cherished images that are printed with the purpose of long-term preservation, the benefits of recording the critical metadata information may well outweigh the disadvantage of the effort involved.

9. Image permanence standards

Clearly, the industry needs standards, both on image permanence and physical durability, as a commonly agreed method to assess the stability of various media. The standards need to be well documented, extensible, and data-driven; the data behind the standards needs to be based on scientifically sound principles. This is especially important as the need for image preservation grows, driving further the need to evaluate media stability and accurately assess tradeoffs. As mentioned previously, preservation requires good information on dark stability. This is more than just thermal stability, which has been the norm in the silver halide world, and includes gas pollutants and humidity as well. The ISO technical committee on photography is currently working on these standards to define methods for testing. Methods for predicting life estimates are also needed, however, and these must include the four environmental factors (those previously mentioned plus light), and they need to be relevant to the specific application and end user. As mentioned, tighter degradation criteria are needed for the application of preservation compared to consumer predictions. The longevity estimates included here are 50% tighter (allow for half the dye loss) than those generally in use for consumer longevity predictions, which are taken from the ANSI Illustrative criteria contained in the silver halide stability standard [7]. These tighter criteria would yield a net change that would be considered close to a just noticeable difference (JND). For this application, that level of change is about the maximum tolerable.

10. Conclusions – A call to action

Much work is being done in academia and in large institutions on the preservation of digital information and images. Concerns around long-term preservation have been recognized in these applications. As digital imaging grows into the mainstream consumer and professional photography space, more education is needed to make the consumer aware of the risks of image loss for not having an archiving and preservation plan. The initiative that was announced in 2007 by the International Imaging Industry Association on consumer photo preservation is a start to creating this awareness among retailers and the consumers [15]. The "SaveMyMemories.org" website, also launched in 2007, has received positive response, and similar websites are in place today. These efforts need support from the entire industry—this includes imaging manufacturers as well as museums and libraries—essentially any institution dealing with images. Creating the awareness will help create the demand for high-volume systems to address the need for image preservation.

This paper has reviewed both traditional (silver halide), and digital (inkjet, thermal) media for storage on both hardcopy prints and film. Electrophotographic media, such as that used for on-line generation of photo albums and scrapbooks also needs to be considered. Recognition is also given to the fact that larger hybrids, as well as all-digital storage systems, are extremely important. All will play a large role in digital preservation depending on accessibility needs and cost constraints. Eastman Kodak Company has played, and will continue to play, a strong role in all of these applications, including entertainment, document, medical, and consumer imaging, but the industry, as a whole, needs to work together to make a viable long-term image preservation system a reality.

11. Acknowledgments

The author acknowledges the efforts of all of the people from the many divisions of Eastman Kodak Company who have been exploring long-term preservation strategies. In addition, he specifically thanks Robert Breslawski, Peter Burns, David Cok, Edward Covannon, Randy Fredlund, Mark Henry, Joseph Manico, Kenneth Parulski, Kenneth Repich, and Donald Williams for their technical help, guidance, and encouragement.

12. Author Biography

Joseph LaBarca joined Eastman Kodak Company in 1976 with a Bachelor's of Science Degree in Chemical Engineering. He has spent the majority of his career in the research, development, and commercialization processes for Kodak Ektacolor papers and processing chemistry in both technical and leadership roles. This included extensive involvement in the stability of color papers beginning in the early 1980s and continuing through today. In 1997, Joe was appointed Senior Research Lab Manager in the Imaging Media and Materials organization of the Kodak Research Labs, directing a laboratory with systems responsibility for professional color negative films, papers, and display materials, and held this position through June of 2004. He is now Technical Director, Image Permanence in the Research Labs of the Consumer Digital Imaging Group with responsibilities that include silver halide, inkjet, thermal, and electrophotographic imaging systems. Joe is member of the American Institute for Conservation. Joe is also a member of the ISO Technical Committee on Photography and is directly involved in the ANSI/IT-9 and ISO Working Group 5 Committee on color print stability.

References

[1] Gschwind R Rosenthaler L and Buchel R 2004 Digitization and Long Term Archival of Photographic Collections: Recommendations of the Swiss Federal Office for Civil Protection, Section Protection of Cultural Property *Proc. IS&T Archiving Conf.* pp 11–17
[2] Kriegsman S and Mandell L. 2004 Digital Archiving Without Preservation is Just Storage: Education is the First Step to Achieving Preservation Goals *Proc. IS&T Archiving Conf.* pp 32–35
[3] Miyata K 2004 Issues and Expectations for Digital Archives in Museums of History: A View from a Japanese Museum *Proc. IS&T Archiving Conf.* pp. 108–111
[4] Wilson A 2005 A Performance Model and Process for Preserving Digital Records for Long-term Access *Proc. IS&T Archiving Conf.* pp. 20–25
[5] Normand C, Gschwind R and Riedel W 2005 Long Term Preservation of Digital Images on Color Microfilm *Proc. 21st Annual IS&T Int. Conf. on Digital Printing Technologies NIP-21* pp 353–356
[6] Hofmann A, Riedel W and Sassenscheid K 2005 ArchiveLaser Project: Accurate Long Term Storage of Analog Originals and Digital Data with Laser Technology on Color Preservation Microfilm *Proc. 2nd Annual IS&T Archiving Conf.* pp 197–200
[7] Stability of Color Photographic Images – Methods for Measuring ANSI IT9.9-1996, and ISO 10977
[8] Bugner D E, Romano C, Campbel G A, Oakland M M, Kapusniak R, Aquino L and Maskasky K 2004 The Technology Behind the New Kodak Ultima Picture Paper – Beautiful Inkjet Prints that Last for Over 1000 Years Proc. Is&t 13th Int. Symp. Of Photofinishing Technologies pp 38–43
[9] Oldfield D and Segur R 2004 Assessment of Current Light-Fade Endpoint Metrics Used in the Determination of Print Life – Part I *J. Imaging Sci. Technol.* **48** 6 pp 495–501
[10] Oldfield R and Twist J 2004 Assessment of Current Light-Fade Endpoint Metrics Used in the Determination of Print Life – Part II *Proc. IS&T Archiving Conf.* pp 36–41
[11] Willaims D A 1996 A Method for Recording and Storing Color Images US Patent Number 5,606,379

[12] Williams D and Burns P 2004 Human-Readable Preservation of Digital Images to Microfilm *Proc. IS&T Archiving Conf.* pp 183–186
[13] Burns P D, Madden T E, Giorgianni E J and Williams D 2005 Migration of Photo CD Image Files *Proc. IS&T Archiving Conf.* pp 253–258
[14] Schewe, J 2006 *About Metadata*, Adobe Systems, Inc. Technical Paper
[15] International Imaging Industry Association 2006 Consumer Photo Preservation Initiative fact sheet distributed at the Photo Marketing Association Convention

Journal of Physics: Conference Series **231** (2010) 012009 doi:10.1088/1742-6596/231/1/012009

Durability of Ink Jet prints

E Dobric[1], I Bolanca Mirkovic[2], Z Bolanca[3]

[1] PhD student, University of Zagreb, Faculty of Graphic Arts, Getaldiceva 2, 10000, Zagreb, Croatia
[2,3] University of Zagreb, Faculty of Graphic Arts, Getaldiceva 2, 10000, Zagreb, Croatia

E-mail: ibolanca@grf.hr

Abstract. The aim of this paper is the result presentation of some optical properties research for ink jet prints after: exposing the prints to the mixed daylight and artificial light, exposing of prints to the sun-light through the glass window, and exposing of prints to outdoor conditions during the summer months.
The prints obtained by piezoelectric and thermal ink jet technologies were used in the researches. The dye-based inks and the pigmented inks based on water and the low solvent inks were used.
The results of these researches, except the scientific contribution in the domain of understanding and explaining the environmental conditions on the gamut size, i.e. the range of color tonality, colorimetric stability and print quality, can be used by the ink and paper manufacturers in new formulations, offer data for the printer producers for further production and evaluation of the position of their products.

1.Introduction

Image quality, durability and storage stability are three major attributes of ink jet prints that attract attention and have driven a continuous improvement in this technology over the past years.

Durability of color prints has steadily improved. Improvements have targeted the following areas: lightfastness, thermostability, water-fastness, humidity-fastness, and pollution gas-fastness [1, 2]. Light fastness is one of the most important items in ink jet printing [3].

Degradation of dyes as it is known depends on a wide variety of environmental parameters like temperature, humidity, light intensity and spectral distribution. Pigment based inks have better resistance to gas and light and better water-fastness than dyes [4]. A lot of researches and empirical explanations known in the literature describe how dye degradation may take place and which the most probable degradation mechanisms are. Previous work with dye- based inks showed that the light fading of cyan and magenta were media dependent. Thermal dye transfer prints have poor image stability when exposed to visible or ultraviolet light or subjected to heat because of the inherent chemical properties of dye molecules [5].

Ozone is known as a very aggressive substance and it belongs to the most powerful known oxidants. Many researches were done to understand the mechanism of the ozone oxidation of dyes [6]. Ozone itself, or the decomposition product singlet oxygen, is believed to be the active species in the oxidation process of organic compounds. Other pollution gases like NO_x and SO_2 are also considered to be aggressive agents, but they seem to play only a minor role in the gas fading phenomenon of colorant [1].

De Rossi and coauthors present research of activation energies on different media with different dyes also considering the influence of room humidity [7].

Journal of Physics: Conference Series **231** (2010) 012009 doi:10.1088/1742-6596/231/1/012009

The purpose of this paper is the result presentation of the gamut research of ink jet prints exposed in the air-conditioned room with mixed daylight and artificial light and exposing the prints to the daylight radiation through the glass window in summer months. The prints obtained by piezoelectric and thermal ink jet technologies were used in the researches. The dye-based inks and the pigmented inks based on water and the low-solvent inks were used. The results of these researches, except the scientific contribution in the domain of understanding and explaining the environmental conditions on the gamut size, i.e. the range of color tonality and print quality, can be used by the ink and paper manufacturers in new formulations, as well as for further production and evaluation of the position of the products.

2. Experimental

Ink jet technology is the printing technology using the ink droplets without the contact with the printing substrate. The ink droplets are sprayed through the nozzles on the substrate to obtain the image.

The prints obtained by piezoelectric and thermal ink jet technologies were used in the research. The following printers were used: Epson Stylus PRO 7000, HP Designjet 9000s and HP Designjet 500.

Epson stylus PRO 7000 has Micro Piezo DX3 printing heads which achieved the high quality prints. Maximal resolution is 1440 x 720 dpi. Six dye based inks (cyan, magenta, yellow, black, light cyan and light magenta) were used.

HP Designjet 9000s uses piezoelectric ink jet technology. The resolution of printing is 720x720 dpi. Six printing heads were used for printing CMYK + C_L + M_L, the same as in the previously described printer. This printer uses carbon filters to absorb the air volatile organic compounds released during the printing process. The system is designed to absorb VOC from air at rates consistent with the print speed and number of compounds released during printing. Low-solvent inks were used for printing.

HP Designjet 500 uses color thermal ink jet technology. Maximal printing resolution is 1200x600 dpi. CMYK water -based ink jet inks were used.

The test form contained ISO and ECI patterns. The part containing the ECI measuring form consists of fields with different combination color values of the subtractive synthesis. It was intended for spectrophotometric analysis. The information quantity obtained by such measurements enabled the construction of 2D and 3D gamut in perceptual uniform color space. ISO illustration was printed for the visual control.

Some of the important paper for ink jet printers properties made are smoothness, gloss and penetration. The mirror reflection of the incoming light appears on the smooth paper surface and the diffuse reflection which is not good for the print quality, appears on the rough surface. Except that, this printing technique demands from the substrate the equableness of the penetration angle in order to prevent the side ink spilling. The coated papers (matt and glossy) are characterized by better mechanical and chemical surface properties which influence the interaction with the ink and its better adherence. Solstar matt paper, declared as the poster paper for indoors and outdoors graphics, was used for printing (substrate1). It can be used in combination with dye-based and pigmented inks, oil, eco/mild solvents.

Many office papers were processed for the usage in ink jet and laser printing technology. Paper (substrate 2) was intended for printing based on electrophotography – laser printing. Because of that, it has special importance in comparison with the substrate 1 in the described experimental conditions.

The prints are divided into three series. In the first series the samples were exposed in the air-conditioned room to the mixed daylight and artificial light in the duration of 1 and 2 months. The samples in the second series were exposed from the inside of the room through the glass window in the duration from 1 to 3 months during the summer period. In the third series prints were exposed to outdoor conditions during the summer months.

After that, the instrumental analysis comprised the measurements made by spectrophotometer, followed by the statistic data processing obtained by measurements of each field. Except that, the conversion from CIEXYZ into CIE L*a*b* system was performed in order to enable the presentation of 3D gamut of samples in the three-dimensional unified color space. The presentations were made at the level L* equals 60. FT-IR spectrophotometer Spectrum One Perkin Elmer was used.

3. Results and discussions

Information distribution depends on different media which have limited information quantity that they can present. Information refers mainly to those concerning the threestimulus value transformation of the determined color. Gamut represents the total range of information on color, i.e. tone, saturation and lightness which can be reproduced by the given medium. Gamut limits present the volumes or surface which is determined by the gamut extremes.

For objective evaluation of the print quality the ECI values of field samples of different combinations of color values of the subtractive synthesis were measured and the values for the construction of 3D color cubic units were obtained by the computer support and conversion.

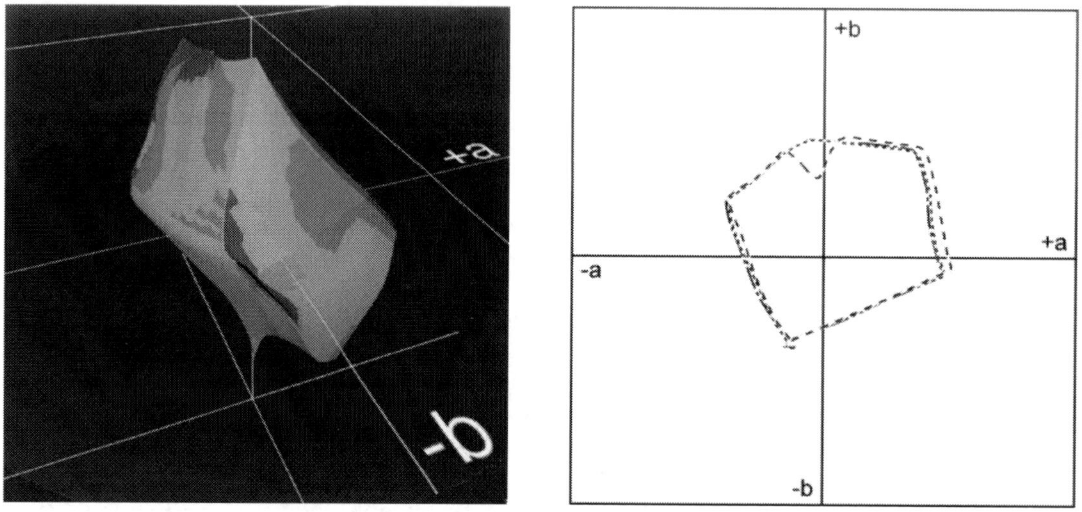

Figure 1. Print gamut 3D and 2D (HP DesignJet 500,sample 1, series 1)

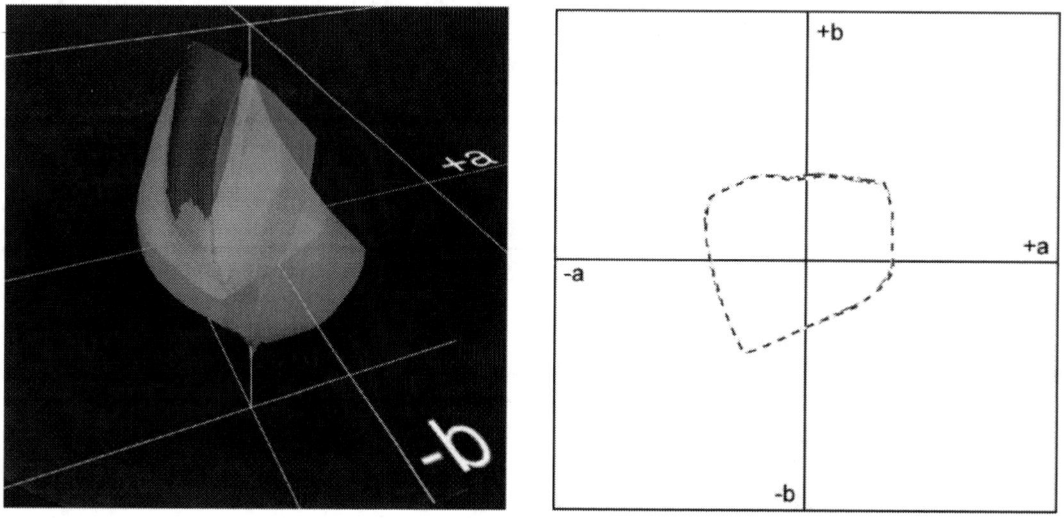

Figure 2. Print gamut 3D and 2D (HP DesignJet 500, sample 2, series 1)

Table 1. Volumes of gamut V CIE L*a*b* CCU (HP DesignJet 500, series 1)

	Before exposure	Exposure 1 month	Exposure 2 months
Sample 1	370553	362984	368766
Sample 2	282024	280554	282082

The samples exposed in the air-conditioned room with the mixed daylight and artificial light in the duration of 1 and 2 months almost did not show the volume change of gamut. The print on the substrate 2 had smaller gamut for 88529 units, which resulted in lower print quality. In sample 1 the gamut changing caused by print ageing were visible in the area of magenta and orange.

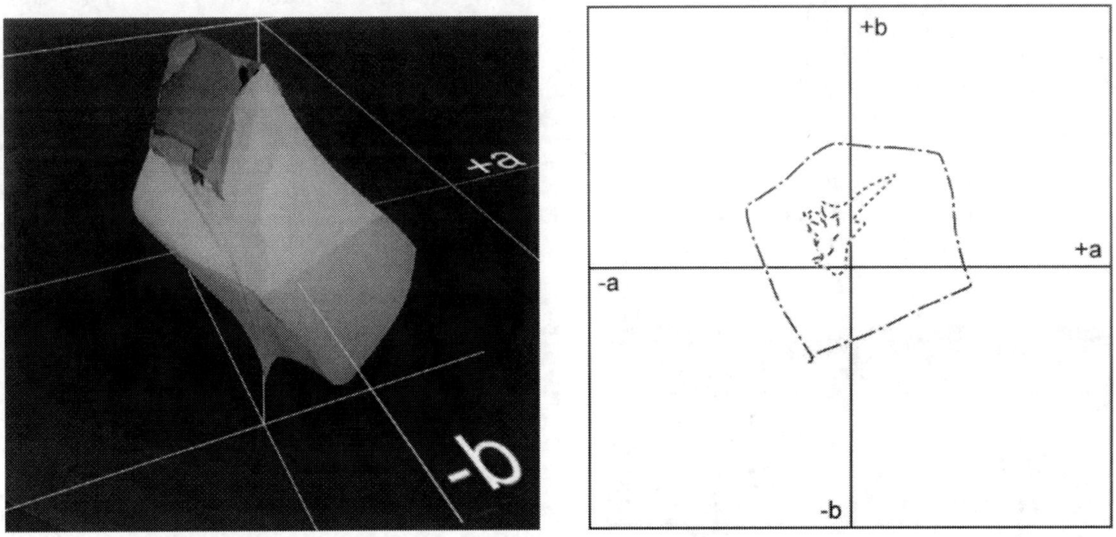

Figure 3. Print gamut 3D and 2D, (HP DesignJet 500, sample 1, series 2)

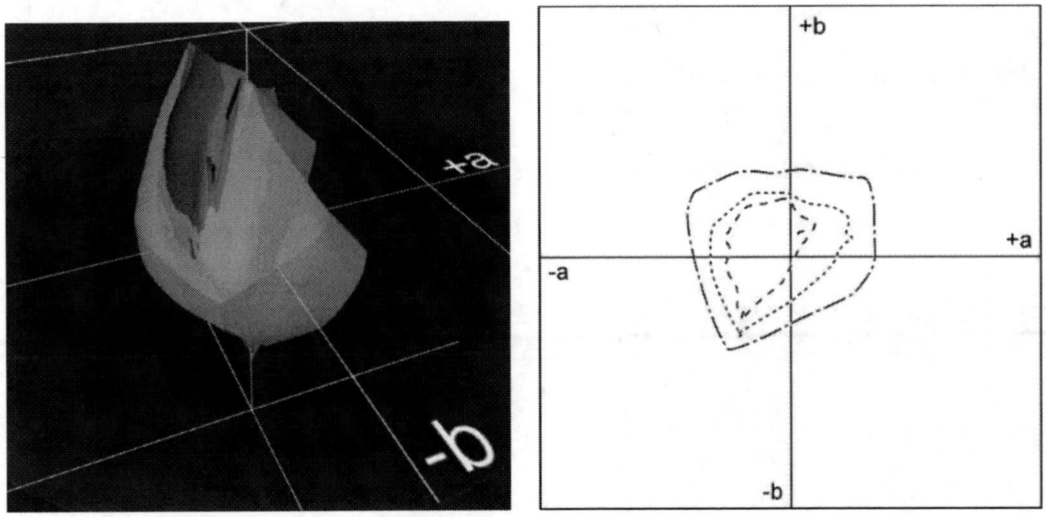

Figure 4. Print gamut 3D and 2D (HP DesignJet 500, sample 2, series 2)

Table 2. Volumes of gamut V CIE L*a*b* CCU (HP DesignJet 500, series 2)

	Before exposure	Exposure 1 month	Exposure 2 months
Sample 1	370553	114717	54848
Sample 2	282024	171098	113186

The samples of the second series were exposed from the inside of the room through the glass window in the duration from 1 to 2 months during the summer period. On the sample 1, after 1 month exposure the gamut volume decreased for 255836 gamut units.

In the second month the influence of print exposure was much smaller and the gamut volume decreased for 59869 units. The decrease of gamut volume was noticeable in all color areas.

When exposing the print through the glass window the gamut volume decrease is noticeable for 110926 units after one month on sample 2 and for further 57912 units after two months. In this case greater changes are visible in the red area.

In figures 5 and 6 the gamuts for prints exposed to outdoor conditions during the summer months are presented. The general weather characteristics were the high air temperatures and sunny weather without greater street level air streaming. From the air pollution source the intensive traffic had to be excluded which pointed at the presence of pollutants, among others oxygen, nitrogen and ozone, which had previously been determined to be the gas fading phenomenon of colorant.

For the print on the substrate 1 the gamut volume decreased for 334352 units, i.e. for 90.23% for exposing it to outer conditions in the duration of one month (figure 5). The trend of gamut decrease continued in the second month but with considerably smaller intensity. The gamut volume decreased for 18419 units or for 49.91%.

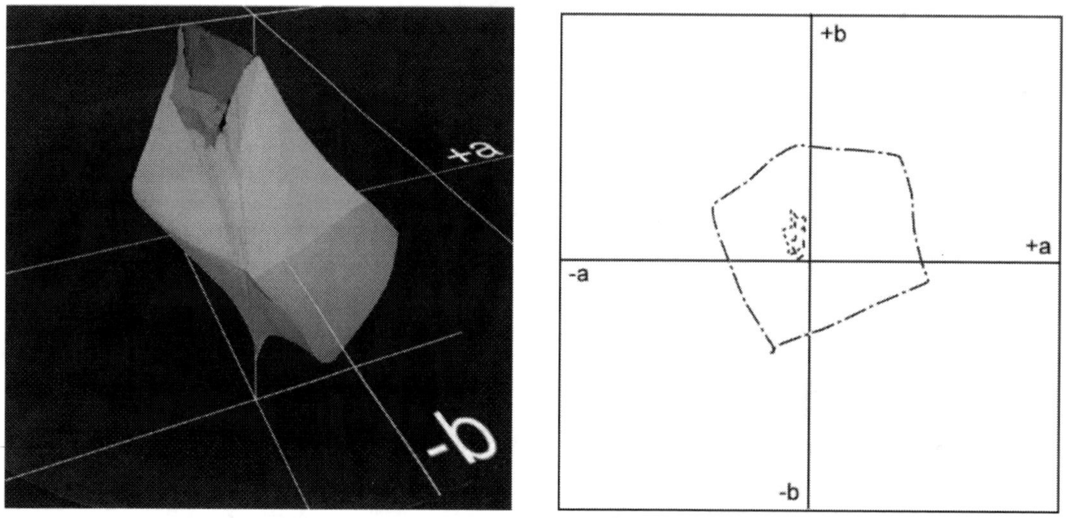

Figure 5. Print gamut 3D and 2D (HP DesignJet 500, sample 1, series 3)

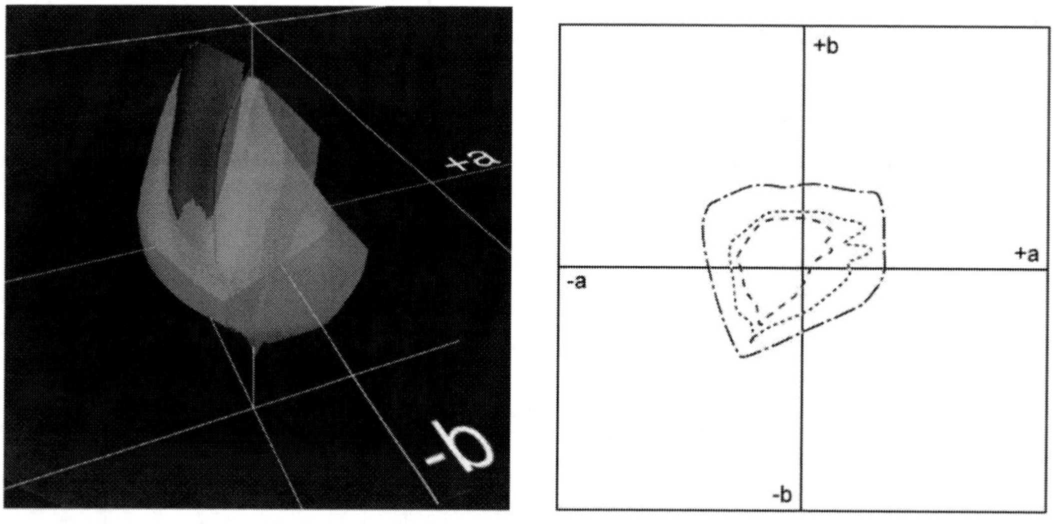

Figure 6. Print gamut 3D and 2D (HP DesignJet 500, sample 2, series 3)

Table 3. Volumes of gamut V CIE L*a*b* CCU (HP DesignJet 500, series 3)

	Before exposure	Exposure 1 month	Exposure 2 months
Sample 1	370553	36201	18419
Sample 2	282024	164485	119850

On the print on the substrate 2 the influence of the environment was even smaller (figure 6). In the first month of exposure the gamut volume decreased for 117539 gamut units or for 41.68%. After exposure in the second month the print gamut decreased for only 27.14%. The results point at the influence size of the ink interaction with the printing substrate on the print stability in the described experimental conditions using the pigment ink for the sample preparation.

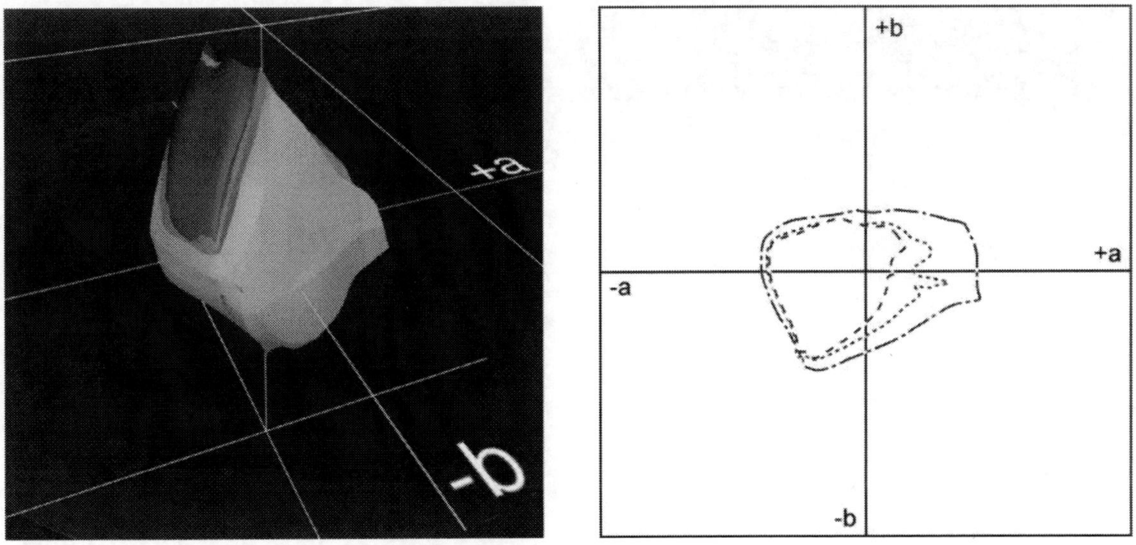

Figure 7. Print gamut 3D and 2D (Epson stylus PRO 7000, sample 2, series 3)

In figure 7, 3D and 2D print gamuts were presented for using the Epson stylus PRO 7000 and six dye based inks under the same criteria of exposure to the outdoor conditions as presented for series 3. In this case far greater durability of print was obtained. After the first month exposure the volume of gamut decreased for 41.68%, and in the second one for 27.14%. The results undoubtedly show that in the application of pigment as well as the dye based inks, in the first month of exposure conditions greater influence, i.e. the prints were less stable regarding the exposition continuation.

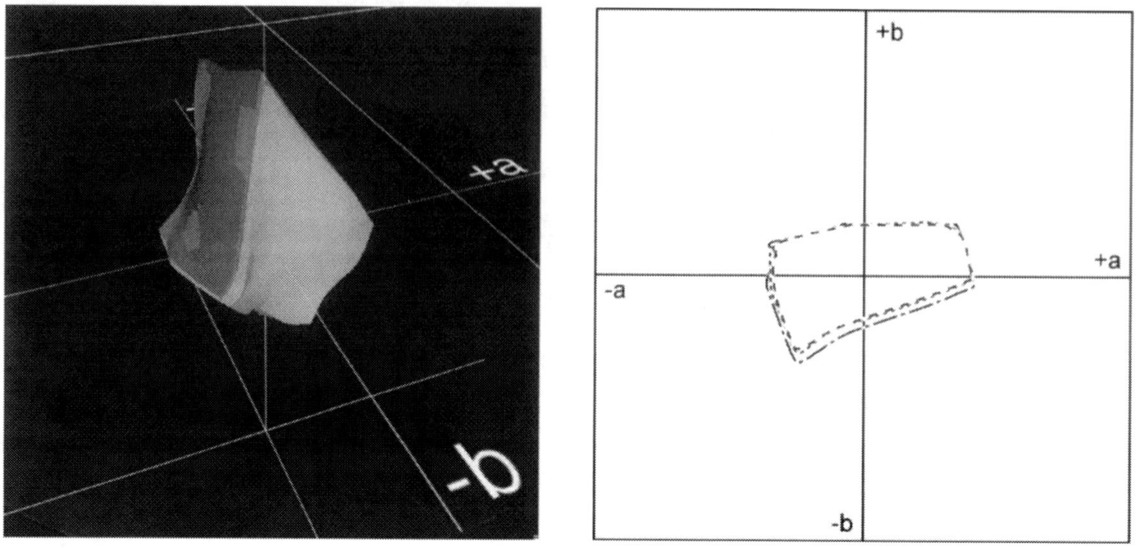

Figure 8. Print gamut 3D and 2D (HP DesignJet 9000s sample 2, series 3)

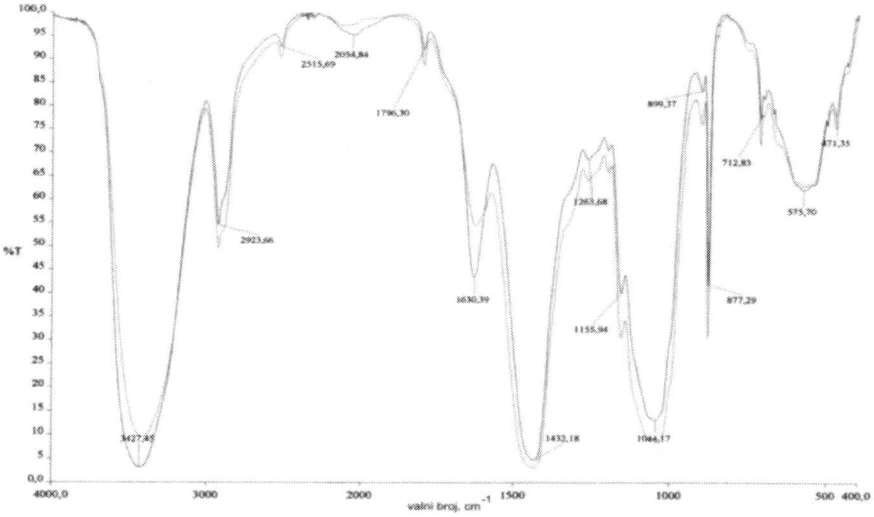

Figure 9. Compared FT-IR spectrum of the sample 2 (unexposed ------, exposed ------)

In the first month of exposition greater influence of the exposure conditions appear for the dye based inks, i.e. the prints are less stable.

In studying of prints stability, it is interesting to include the stability of the print and the substrate. In figure 9, the comparative FT-IR spectra of the sample 2 before and after exposure are presented. The change in the intensity of the characteristic peaks of these two spectra is obvious. The intensity difference of planar vibration of the connection which can be caused by the exposure conditions is noticeable. Somewhat greater absorption in the area from 1200 – 1000 cm^{-1} characteristic for C-O stretching appeared.

However, for explanation of the problems the information which originates from the interaction of the substrate and the determined ink in the function of the exposition conditions is important. The reflection curves in the area of wave lengths from 390 – 700 nm for the unexposed and the exposed print using the HP DesignJet 500 and Epson stylus PRO 7000 for sample 2 and series 3 (figure 10) are presented.

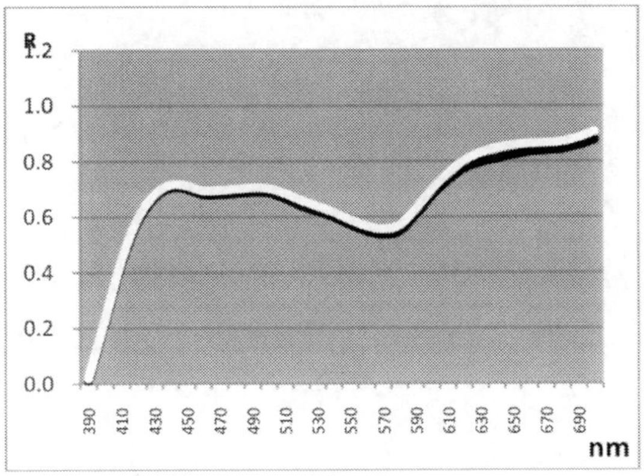

20% screen value

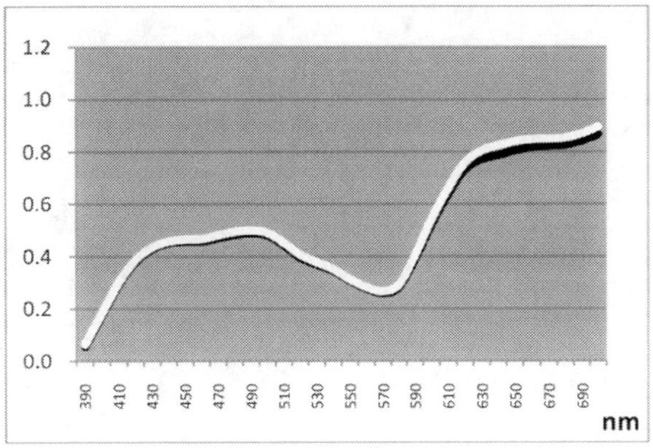

50% screen value

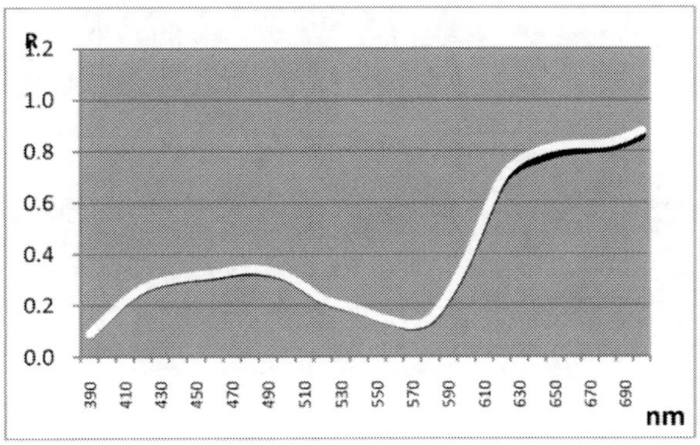

80% screen value

HP DesignJet 500, sample 2, series 1

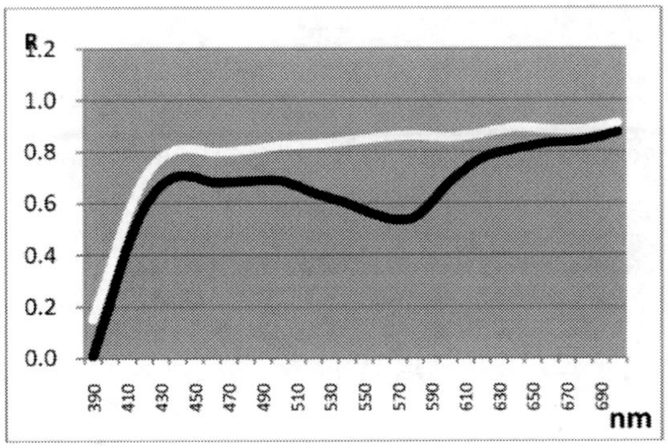

20% screen value

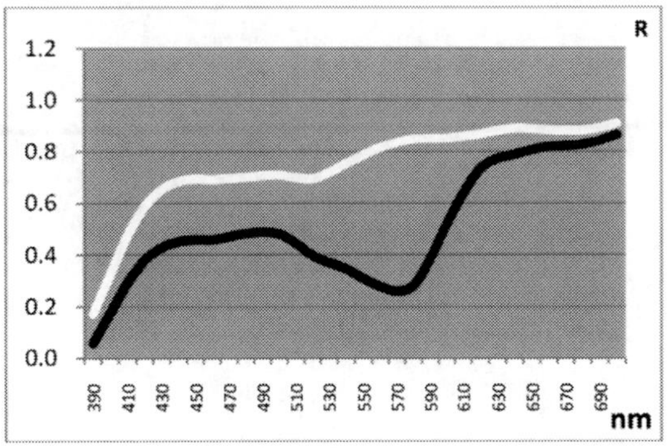

50% screen value

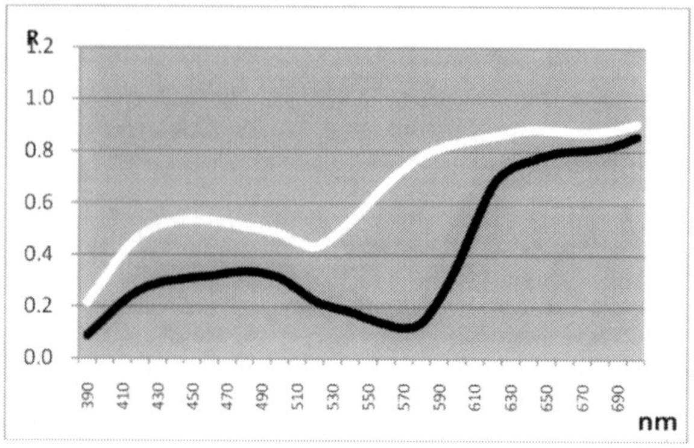

80% screen value

HP DesignJet 500, sample 2, series 3)

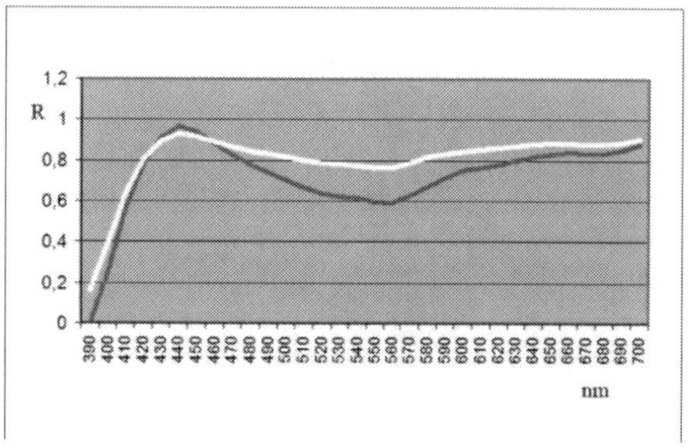

20% screen value

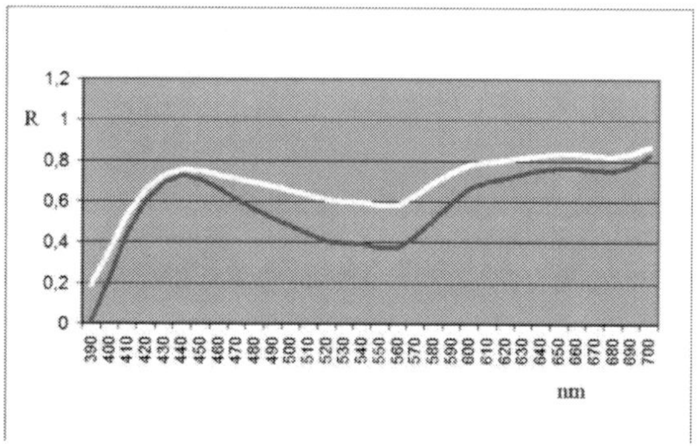

50% screen value

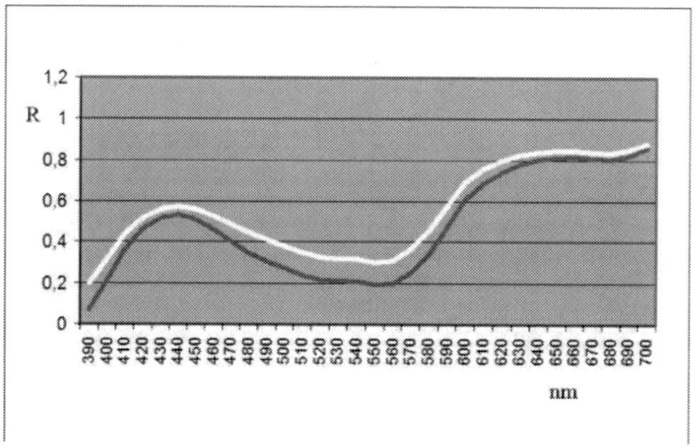

80% screen value

Epson stylus PRO 7000, sample 2, series 3

Figure 10. Reflectance spectra of the prints with magenta (20%, 50% and 80% screen value)

In reflectance curves for 20% screen value the reflection was increased in the area of shorter wave lengths (violet – blue spectrum part) which was caused by the addition of optical whitening agents in the printing substrate. By the increase of screen value, the influence of the substrate was slowly decreased and the rapid fading of magenta was noticed by exposing the sample to the outdoor conditions where the kind of inks in combination with the printing substrate became prominent.

The obtained results of the researches, presented here only in one segment, could be explained by the mechanisms which include photooxidation/reduction of the chromophores. Such mechanisms are especially important in the case of azo colorants. The rapid shift of adsorption peak at 560 nm could be noticed in magenta in the presented spectrum. This is the reason of colorant modification.

4. Conclusion

Based on the results the following conclusion can be made:

- On prints obtained with the printer HP DesignJet 500, the exposition in the air-conditioned room to the mixed daylight and artificial light in the duration of 1 and 2 months almost does not influence the volume of gamut .
- Print on the substrate 2 has smaller gamut for 23.89 units in relation to the sample 1. By exposing the print on the substrate 1 through the glass window in the duration of 1 month during the summer period the gamut is decreased for 69.04%, and for the print on the substrate 2 it decreases for 39.33%. By prolonging the exposition period for another month the influence of the exposition conditions on the volume of gamut is lesser for about 14-24% depending on the substrate.
- For prints exposed to outdoor conditions during the summer months in the duration of one month, gamut is decreased from 49. 91 to 90.23% depending on the used printing substrate.
- Far greater stability of print is obtaine by using the dye based inks in relation to the pigments in combination with the printing substrate.
- From the reflectance spectra of the prints with magenta (20%, 50% and 80% screen value) the rapid shift of the absorption peak at 560 nm can be noticed which influences the modification of colorant and which is explained by the mechanism of chromophores photooxidation.

Except the scientific contribution in the domain of understanding and explanation of the environment conditions and the size of the print gamut which reflects on their quality the investigation results have the applicative character. The application of the results can be in the area of new formulation of the graphic materials in the area of InkJet technique as well as in the development of the test methods for the influence of dry time on performance of InkJet print, and the prints stability, which will be the continuation of these investigations.

References

[1] Hayashi H Kitamura K and Oki Y 2002 Gas-fastness of Photographic Prints of Epson PM-950C *Proceedings of ICIS 02* Tokyo pp 539-542

[2] Graczyk T and Xie B 2000 Lamination Study of Ink Jet Media *Tappi J.*, 83, 6, pp 63-72

[3] Vikmana K Iittia H Matousek P Towrieb M Parkerb AW and Vuorinena T 2005 Kerr gated Resonance Raman Spectroscopy in Light Fastness Studies of Ink Jet Prints *Vibrational Spectroscopy* 37 (c) pp 123–131

[4] Madaras M Diehl D Link S Eiff S and DiCillo J 2008 Polimeric Dye Inkjet Colorants with High Waterfastness *Proceedings of NIP24th and Digital Fabrication* pp 355-358

[5] Abe T 2003 Materials for Providing High Print Quality and High Image Stability *Proceedings of 19th International Conference on Digital Printing Technologies* IS&T pp 390-393

[6] Wight P 2002 Features of Reciprocity Failure and Ozone Fading in Ink Jet Images *Proceedings of 18th International Conference on Digital Printing Technologies* IS&T San Diego, pp 334-336

[7] De Rossi U Litz U and Blendl C 2003 Activation Energy of Dye Degradation Process on Different Ink Jet Papers *Proceedings of 19th International Conference on Digital Printing Technologies* pp 450-453

Journal of Physics: Conference Series **231** (2010) 012010 doi:10.1088/1742-6596/231/1/012010

Effect of fluorescent agents on the permanence of printed materials

Swati Bandyopadhyay[1] **and Phil Green**[2]

[1]Printing Engineering Department, Jadavpur University, Salt Lake Campus, Block-LB, Plot-8, Sector- III, Salt Lake, Kolkata-700098, West Bengal, India

[2]London College of Communication, Elephant and Castle, London SE1 6SB, UK

swatib1@yahoo.com

Abstract. The decay of fluorescence on a range of commercially available papers was investigated by exposing them to a 1KW tungsten lamp for a total of 10 hours, with the samples removed for measurement at regular intervals. The results indicated that FWA decay is more rapid on papers with higher FWA efficacy. Only the papers with minimal FWA efficacy reached a plateau after which there was no further significant decay.

1. Introduction

Most commercially available papers have fluorescent whitening agents (FWAs) added to increase brightness and whiteness. FWAs used in papermaking are stillbene molecules that are excited by radiation in the ultra-violet and emit light in the visible region. From a conservation perspective, these FWAs lead to two problems: firstly, the measurement and viewing of such materials is dependent on the UV content of the illumination used; and secondly, permanence is low owing to the tendency of the FWA efficacy to reduce over time, primarily on exposure to UV radiation. These problems are exacerbated by the different illumination sources standardised by the papermaking and printing industries for the measurement of paper and prints, and further by a divergence between the standard daylight viewing conditions used in visual evaluation of papers and prints, and the actual UV content of end use conditions.

ISO 13655 [1] and 3664 [2] (defining graphic arts measurement and viewing respectively) have recently been revised in an attempt to address these problems by requiring stricter conformance with the specified spectral power distribution.

To investigate this problem, a range of commercially available papers were subjected to a 1KW tungsten lamp in a Fadeometer for a total of 10 hours, with the samples removed for measurement at regular intervals.

One issue encountered is that fluorescent efficacy is not well defined. One widely-used method is to calculate the fluorescent component of CIE Whiteness, but this relates to colorimetric rather than spectral quantities. Loeffler and Green [3, 4] recently proposed a spectral method based on summing the differences between a UV included and UV excluded measurement.

Print permanence relates to the appearance of a print with time. The ability of a print to maintain accurate color strength with time due to light exposure and subsequent fading is very important since the demand of printing industry to maintain color precision and print permanence is high. The rate at which deterioration occurs depends primarily on two main factors: the print itself, that is, the ink and the substrate, and the environmental condition the print is exposed to.

Fluorescent whitening agents (FWAs), also known as optical brightening agent (OBAs), are added to most commercial papers to increase the apparent whiteness or brightness of papers as they absorb ultraviolet radiation and re-emit it in the blue region of the visible spectrum. This can also compensate for a yellow tint of many types of pulps. The derivatives of diaminostilbene sulfonates are used as FWA that are highly conjugated molecules having planar structure and anionic charge. There are different grades of FWA, but papermakers prefer tetrasulfonated stilbene as it has intermediate solubility in water and it is readily retained on fibers, especially if alum or another cationic material is present.

2. Fluorescent efficacy and fading

In this study,

$$Fluorescent\ efficacy = \frac{R - R_{uvcut}}{R} \quad (1)$$

where R and R_{uvcut} are the sum of the spectral reflectance factors of a paper, measured without and with a UVcut filter respectively.

Fading(at a particular time interval) is defined as follows:

$$Fading = \frac{R - R_t}{R} \quad (2)$$

where

R = Sum of initial spectral reflectance factors without UV cut filter at time t=0, and:

R_t = Sum of spectral reflectance factors without UV cut filter at time t

3. Experiment

A range of commercially available papers was used in this study. The fluorescent whitening agent (FWA) content of the papers were measured according to equation (1). The instruments used for this purpose were a Gretag-Macbeth Spectrolino and a Datacolor Spectraflash. The spectral reflectances were measured with and without a UVcut filter. The (initial) FWA content of the different papers is given in Table 1.

Table 1. FWA content of the papers tested

Paper grade	FWA content
1	0.011
2	0.006
3	0.053
4	0.103
5	0.089
6	0.040
7	0.004

The paper samples were placed inside a Fadeometer and exposed to a 1KW tungsten lamp for a total of 10 hours, with the samples removed for measurement at regular intervals. The Fadeometer has water cooling but does not have control of humidity or temperature.

5. Result and Discussion

The different types of papers in this study have different levels of fluorescent whitening agents. The fluorescent efficacy of papers were defined as described by Loeffler and Green [3, 4].

The results of fading of the different papers is given in Figures 1-5.

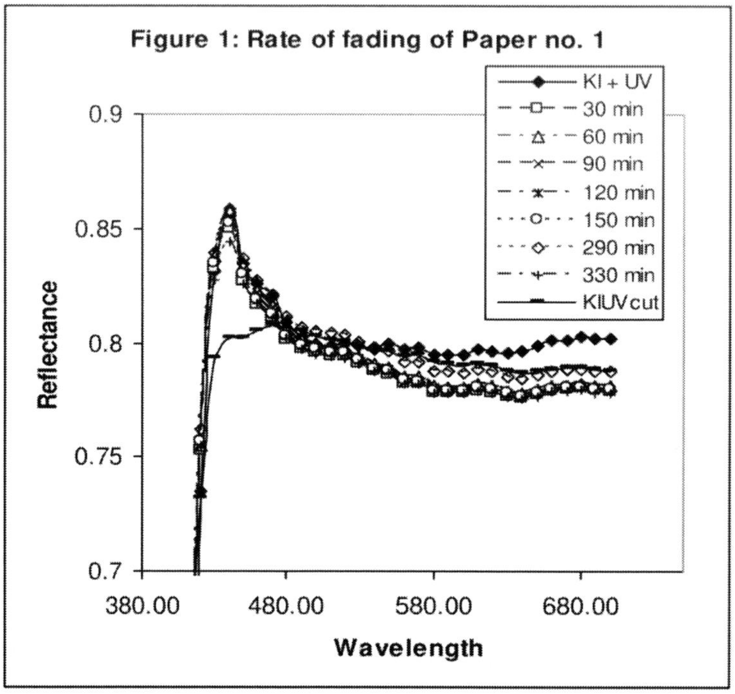

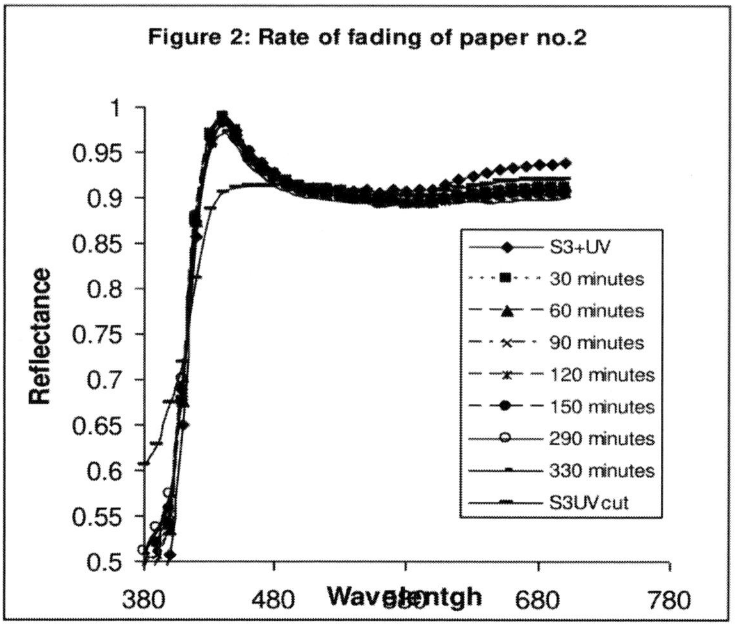

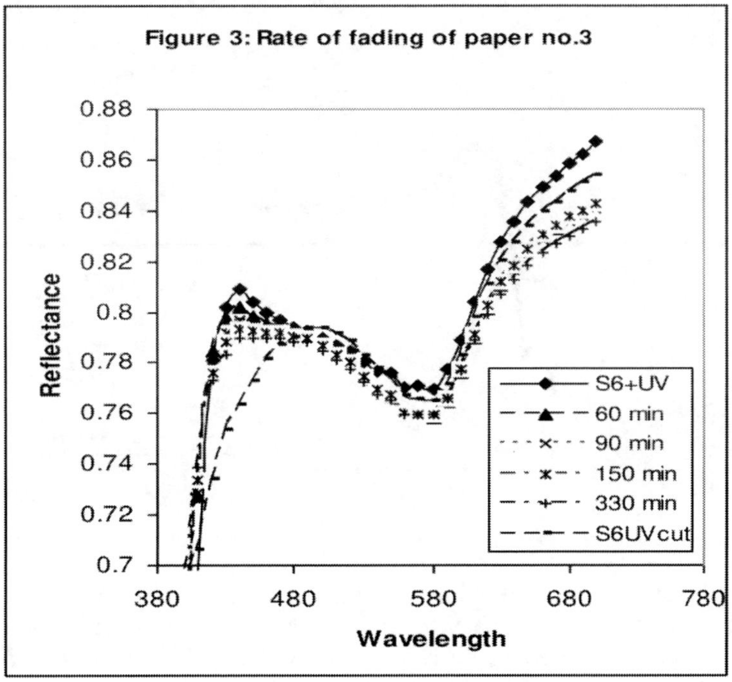
Figure 3: Rate of fading of paper no.3
0.88
0.86
0.84
0.82
0.8
0.78
0.76
0.74
0.72
0.7
Reflectance
380
480
580
680
780
Wavelength
S6+UV
60 min
90 min
150 min
330 min
S6UVcut

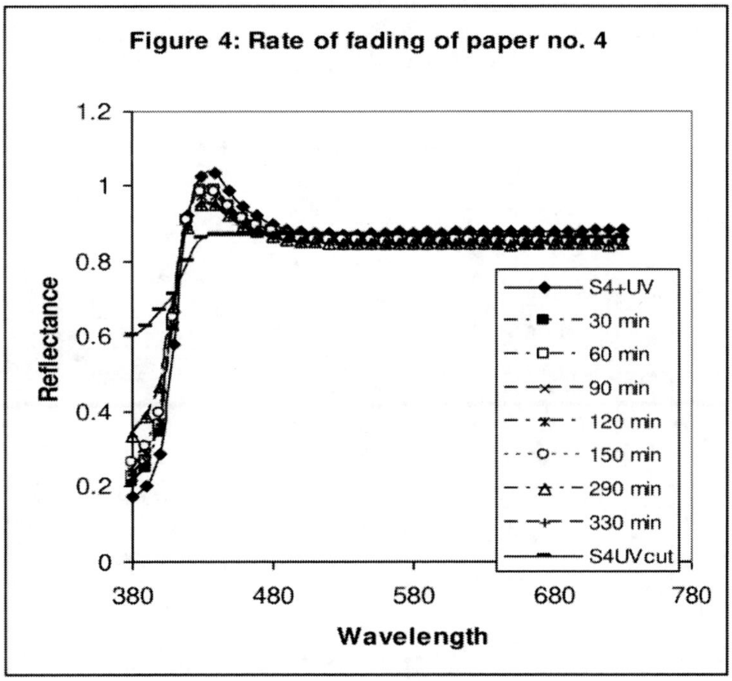
Figure 4: Rate of fading of paper no. 4
1.2
1
0.8
0.6
0.4
0.2
0
Reflectance
380
480
580
680
780
Wavelength
S4+UV
30 min
60 min
90 min
120 min
150 min
290 min
330 min
S4UVcut

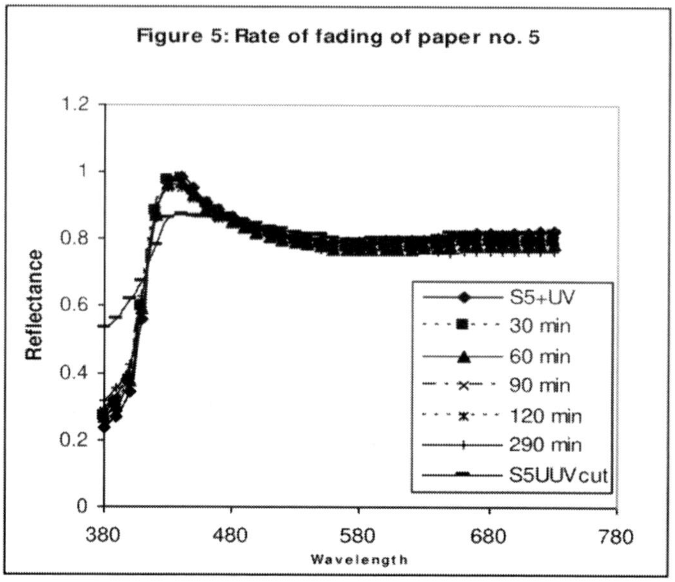

It can be observed that the degree of fading depends on the FWA content of paper, and that the rate of fading is almost negligible for papers having little FWA content. The degree of fading against time for different types of papers is plotted, and it can be observed that the extent of fading varies with time in a power series with a negative coefficient. The rate of fading is initially high and decreases with time.

The power series coefficients of rate of fading are shown against the FWA content in figure 6. A linear correlation ($R^2 = 0.8577$) indicates that the rate of fading rises with increasing FWA content.

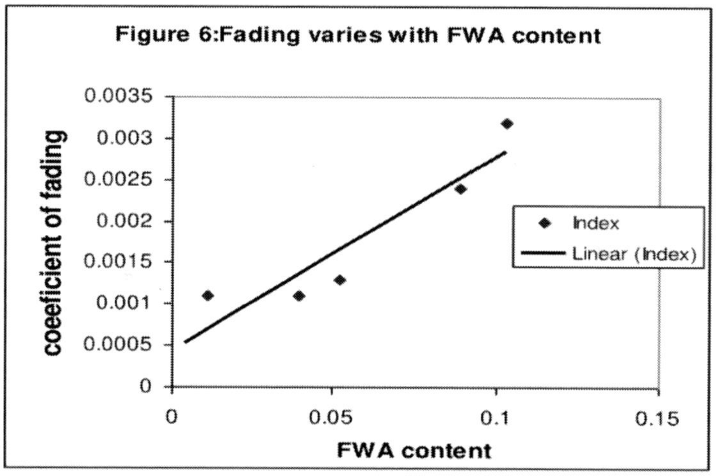

References

[1] ISO 13655:2009 Graphic technology and photography — Spectral measurement and colorimetric computation for graphic arts images, ISO, Geneva

[2] ISO 3664:2009 Graphic technology and photography — Viewing conditions, ISO, Geneva

[3] Loeffler E and Green P J 2008 Estimating the Spectral Reflectance of Fluorescent Offset Papers for Varying Illuminants *Proc. IARIGAI 2008*

[4] Loeffler E and Green P J 2008 Scalable Model Spectra Enabling Estimation of Fluorescence Characteristics of Brightened Offset Papers Under Different Illuminants *Proc ISCC Special Topics Meeting 2008*

Journal of Physics: Conference Series **231** (2010) 012011 doi:10.1088/1742-6596/231/1/012011

Crystal violet: study of the photo-fading of an early synthetic dye in aqueous solution and on paper with HPLC-PDA, LC-MS and FORS

Daria Confortin[a], Han Neevel[b], Marina Brustolon[a], Lorenzo Franco[a], Albert J. Kettelarij[c], Renè M. Williams[c] and Maarten R. van Bommel[b].

[a] Dipartimento di Scienze Chimiche, Università degli Studi di Padova, via Marzolo 1, 35131 Padova, Italy

[b]Netherlands Institute for Cultural Heritage (ICN), P.O. Box 76709, 1070 KA, Amsterdam, The Netherlands

[c] Molecular Photonics Group, Van't Hoff Institute for Molecular Sciences, Faculty of Science, Universiteit van Amsterdam, Nieuwe Achtergracht 129, 1018 WS Amsterdam, The Netherlands

corresponding author's e-mail address: daria.confortin@gmail.com

Abstract. The photo-fading of crystal violet (CV), one of the earliest synthetic dyes and an ink component, is examined both in solution and on paper. Aqueous solutions of CV were exposed to UV light (365nm) and samples were taken at constant time intervals and analysed with a High Performance Liquid Chromatography-Photo Diode Array (HPLC-PDA) and Liquid Chromatography-Mass Spectroscopy (LC-MS). Demethylation products were positively identified. Also, deamination probably occurred. The oxidation at the central carbon likely generates Michler's ketone (MK) or its derivatives, but still needs confirmation. To study CV on paper, Whatman paper was immersed in CV and exposed to UV light. Before and after different irradiation periods, reflectance spectra were recorded with Fibre Optic Reflectance Spectrophotometry (FORS). A decrease in CV concentration and a change in aggregation type for CV molecules upon irradiation was observed. Colorimetric L*a*b* values before and during irradiation were also measured. Also, CV was extracted from paper before and after different irradiation periods and analysed with HPLC-PDA. Photo-fading of CV on paper produced the same products as in solution, at least within the first 100 hours of irradiation. Finally, a photo-fading of CV in the presence of MK on Whatman paper was performed. It was demonstrated that MK both accelerates CV degradation and is consumed during the reaction. The degradation pathway identified in this work is suitable for explaining the photo/fading of other dyes belonging to the triarylmethane group.

1. Introduction

Since the XIX century, synthetic dyes have been used as colouring matter in many inks. Triarylmethane dyes, part of a group trivially named aniline dyes, were among the first ones to be produced and marketed, Mauvein being the first synthetic aniline dye to be synthesized in 1856 by William Perkin. Due to their high colouring power and relatively inexpensive production methods,

Journal of Physics: Conference Series **231** (2010) 012011 doi:10.1088/1742-6596/231/1/012011

these aniline dyes initially gained great popularity, but soon their bad light fastness properties became apparent. In particular, Crystal violet (CV) or hexamethyl pararosaniline (figure 1A) with its deep purple hue has had and still has a widespread diffusion for dyeing textiles and paper, as an ingredient of writing and drawing inks and, as a pigment, consisting of the copper ferrocyanide lake, in printing inks [1]. Methyl Violet (MV), which consists of a mixture of tetra-, penta- and hexa- methylated pararosaniline was invented in 1861 and introduced to the market in 1866. An efficient synthetic route to CV was only developed in 1883.

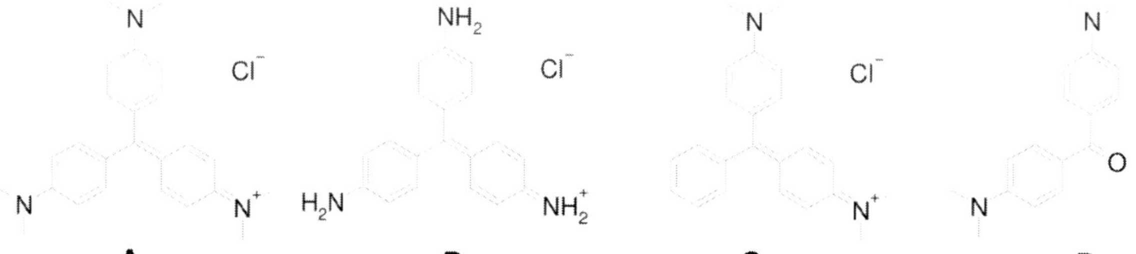

Figure 1: Structures of Crystal Violet or hexamethyl pararosaniline (A), pararosaniline (B), diamond green (or basic green 4) (C) and Michler's ketone (D).

Since then, discoloration has shown to be a crucial issue. Indeed, ink fading, leading to chromatic alterations of documents or drawings or even to a complete loss of information, was soon observed.

In particular, among Van Gogh's signed works there is a group of drawings and letters produced in 1888 in Arles and all made with a purple ink which was demonstrated to contain methyl violet, PR and other methylated derivatives whose structures are not known yet. Interestingly, a seemingly brownish drawing from the same group, entitled Montmajour (Arles, 1888), shows purple shades on the edges where the ink has been protected from light by the frame [3] (figure 2).

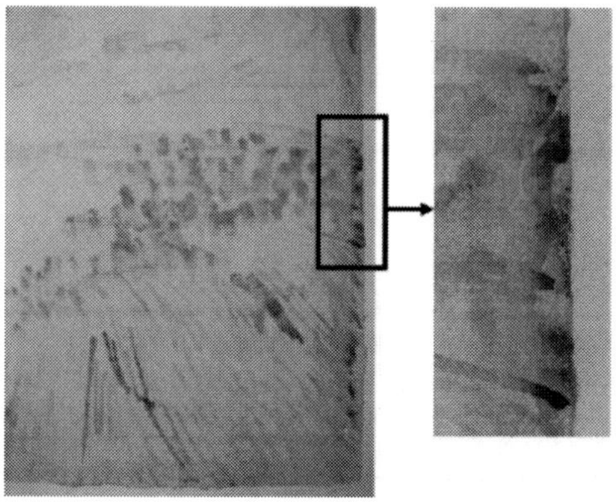

Figure 2: On the left: detail of Montmajour drawing (Van Gogh, Arles, 1888). On the right: enlarged view showing different purple shades (photo reproduced with permission by Van Gogh Museum -Amsterdam).

Consequently, the question of whether other drawings and letters by Van Gogh which are now brown-coloured were originally purple, needs to be answered.

Therefore the knowledge of synthetic dyes degradation mechanisms and the possible identification of original dyes from their degradation products in historical samples are gaining more and more importance in the field of cultural heritage study and conservation.

CV shows poor light fastness, in particular on paper and on natural fibers such as cotton, silk and wool. Moreover these substrates are not inert to light, and the reactivity of the dye-substrate system is complicated by the degradation of the substrate when exposed to light. For this reason, many studies have dealt with the reactivity of the dye in solution (aqueous solution or solutions mimicking the substrates) in an attempt to simplify the complex reactivity of the dye-substrate system. Of course, when dealing with a real object such as a drawing or a print, it is not possible to neglect the substrate reactivity [4] and a study of model systems for the dye on paper is necessary [5].

The objective of this work is to study the photo-fading of CV via HPLC-PDA, LC-MS and FORS both in aqueous solution and on paper in order to use the information obtained from the former system to best interpret the results of the study of the degradation of the dye on paper. The final goals are:

- to determine the composition of the purple ink used by Van Gogh;
- to shed light on the discoloration mechanism of purple drawings and letters by Van Gogh and contemporaries;
- to obtain information on the original colour of discoloured inks on paper.

2. Experimental

2.1. Sample preparation

Two aqueous solutions of CV (Acros) 5×10^{-5} M were prepared using ultrapure water (Simplicity system Millipore). Afterwards, both solutions were exposed to UV light (365 nm) from a Spectroline® Super High Intensity black light lamp (SB-100/F Spectronics Corporation, USA) for at least 100 hours and samples were periodically taken.

CV was applied to Whatman paper discs (Cat No 1003 125) by soaking them for 10 minutes in an aqueous solution of CV (5×10^{-4}M) or in an ethanol solution of CV (5×10^{-4}M) and MK (1×10^{-3}M). Afterwards, the discs were left to dry horizontally.

Samples (1x0.5 cm strips of paper) were analyzed before and after different UV irradiation intervals. The dye was extracted from each sample with 200 µl methanol (Fluka, for HPLC) at 70 °C for 10 minutes. Samples of un-dyed paper were treated in the same way and used as a reference.

Pararosaniline (Acros) was used as received and Diamond Green B was obtained from the Dyes Reference Collection of ICN.

2.2. HPLC-PDA analysis

HPLC analysis was performed with equipment from Waters Chromatography BV (Etten-Leur, The Netherlands). Mobile phase was delivered at a flow rate of 0.2 mL/min by a 616 LC pump, controlled by a 600S controller. An in-line degasser degassed all solvents used. Samples were injected by a 717 autosampler. Detection was performed with a 996 Photo Diode Array (PDA) detector equipped with a 10 µL detector cell. The equipment was controlled by a computer with Empower software from Waters Chromatography BV; the same system was used for data acquisition. Separation was performed on a Luna C18 (2) column (150mm x 2mm id) protected by a security C18 guard column, both supplied by Phenomenex (Torrance, CA, USA). Mobile phase consists of a gradient of water (purified by a Simplicity system Millipore, Amsterdam, The Netherlands), methanol (Fluka, Zwijndrecht, The Netherlands) and 5% phosphoric acid in water (ACS reagent, Sigma, Zwijndrecht, The Netherlands). The composition of the solvents and the gradient profile is given in Table 1. Identification is based on both PDA spectra and retention time of the components which are compared with those of known reference materials.

Table 1. Gradient profile for the HPLC system.

Time/min	%A[1]	%B[2]	%C[3]
0	74	16	10
15	0	90	10
20	0	90	10
23	0	100	0
27	0	100	0
30	74	16	10
50	74	16	10

[1] A=10% methanol/water (v/v)

[2] B=100% methanol
[3] C=5% phosphoric acid (w/v)

2.3. LC-MS analysis
LC-MS analysis was performed with equipment from Thermo Electronic corporation. Mobile phase was delivered at a flow rate of 1 mL/min by a Finnigan surveyor LC Pump plus. Samples were injected by a Finnigan surveyor Autosampler plus. Detection was performed with a Finnigan surveyor PDA Plus detector. Separation was performed on a Intersil ODS-3 column (4.6 × 50mm). The mobile phase consists of a gradient of water, formic acid and acetonitrile. Mass analysis was performed with the use of a Finnigan LXQ Ion Trap apparatus.

2.4. FORS analysis
FORS analysis was performed with a home-assembled equipment composed of a halogen lamp (AvaLight-Hal, Avantes), a reflection integrating sphere (ISP-50-8-REFL, Avantes) and a spectrophotometer (AvaSpec-2048, Avantes). The equipment was controlled by AvaSoft 6.1 software. Reflectance spectra were collected in the range 380-800 nm, with an integration time of 200 ms and averaging 5 scans. Absorption spectra were calculated from reflectance spectra using Kubelka-Munk transform. Reflectance spectra for CV were recorded before and after exposure to UV light of a dyed disc of paper. Colorimetric L*a*b* values were calculated from reflectance spectra for a D65 illuminant and a 10° Standard Observer.

3. Results and discussion

3.1. CV in aqueous solutions

3.1.1 *HPLC-PDA and LC-MS analysis.* A colour shift from purple to reddish was observed upon irradiation of an aqueous solution of CV. The HPLC chromatograms for CV in aqueous solution before and after prolonged irradiation (115 hours) are shown in figure 3.

Before irradiation only two peaks are visible at around 21.1 min (relative area >98%) and 20.7 min (relative area <1%), due to CV and mono-demethylated CV respectively (as it will be demonstrated later).

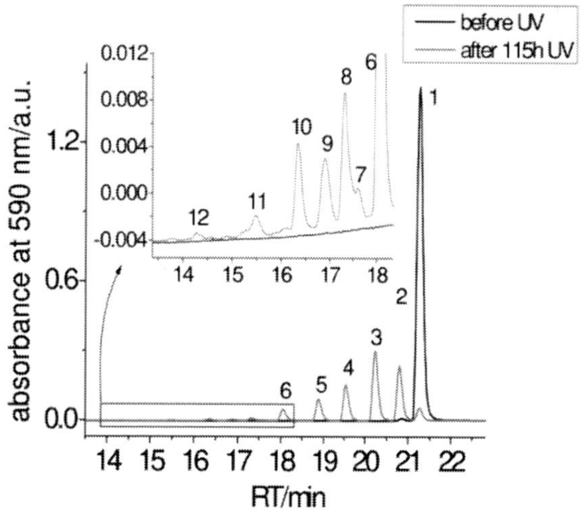

Figure 3: Chromatograms of CV(aq), before and after prolonged irradiation.

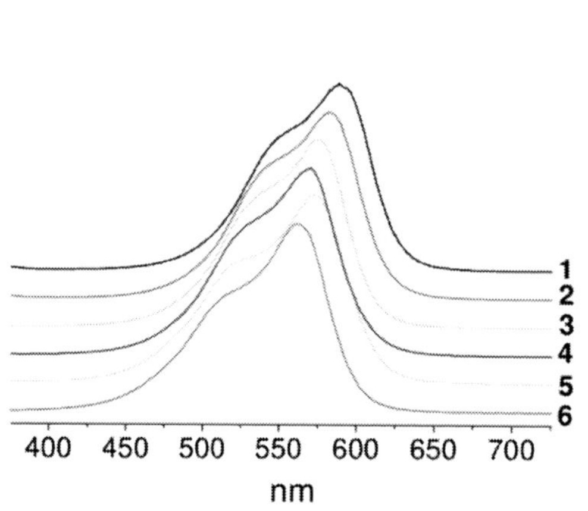

Figure 4: Normalized absorption spectra for peaks 1 to 6 in the chromatogram of irradiated CV(aq) .

Upon irradiation, a decrease of the CV concentration and the formation of at least 11 degradation products were observed.

Peaks 2, 3, 4 and 5 were identified by LC-MS analysis as mono-, bis-, tri-, tetra- demethylated CV. Incidentally, mono-demethylated CV is penta-methylated pararosaniline, bis-demethylated CV is tetra-methylated pararosaniline, tri-demethylated CV is tri-methylated pararosaniline and tetra-demethylated CV is bis-methylated pararosaniline. Absorption spectra for CV and for HPLC peaks 2,3,4,5 are shown in figure 4 together with the spectrum of peak 6 that is believed to be penta-demethylated CV. Demethylation causes a blue shift in the absorption spectra (with the exception of peak 5) and a decrease in retention time, the latter effect being probably due to an increased polarity of the molecules formed. LC-MS analysis revealed the presence of two isomers both for bis- and tri-demethylated CV.

A comparison between the absorption spectra of pararosaniline (RT=14.07 min; λmax=543nm) and of peak 12 in the chromatogram of CV after 115 hours of irradiation (figure 5) enabled to positively identify peak 12 as fully-demethylated CV (pararosaniline). The absorption spectrum of HPLC peak 10 looks very similar to that of pararosaniline, although its maximum has shifted 11 nm to the red (figure 5).

The absorption spectra of peaks 7, 9 and 11 might be due to the products of deamination reaction, such as diamond green which was detected in faded samples of CV [6]. Peaks 7, 9 and 11 have very similar absorption spectra (figure 6), the main difference being a blue shift in the absorption maximum in going from peak 7 to 11 as it was observed for the different products of CV demethylation. Moreover, absorption spectrum for peak 7 is similar to that of diamond green B (figure 1C) although there is no retention time matching. All this suggests the possible formation of various demethylated diamond green-like compounds during CV degradation.

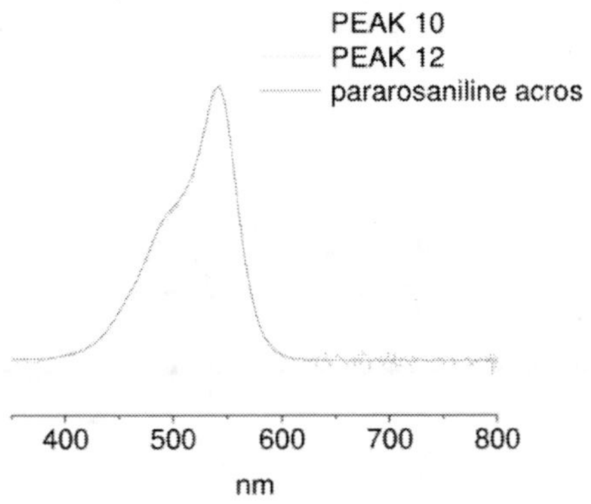

Figure 5: Normalized absorption spectra for pararosaniline and for peaks 10 and 12 in the chromatogram of CV(aq) after irradiation.

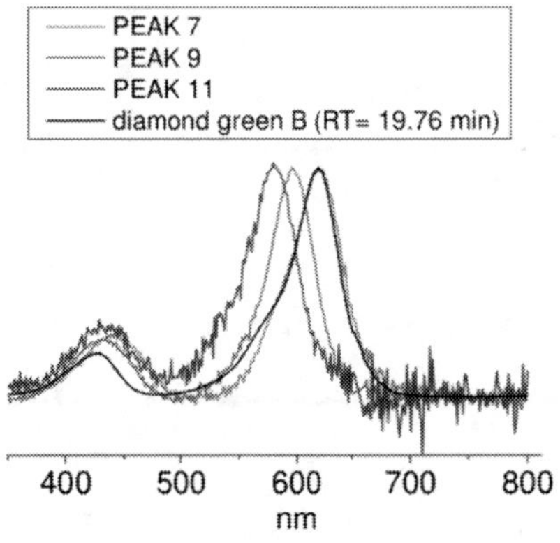

Figure 6: Normalized absorption spectra for peaks 7, 9 and 11 in the chromatogram for irradiated CV(aq) and for diamond green B.

Another hypothesis for the identification of peaks 7, 9 and 11 is reported by Duxbury [7]: in this work it was reported that the absorption spectra of N-oxide derivatives of CV were very similar to the spectrum of diamond green. Moreover, the higher polarity for N-oxides would explain the lower retention times obtained for peaks 7, 9 and 11 with respect to diamond green.

Finally, the absorption spectrum of peak 8 (not shown) has not been identified yet but it is clearly due to the superimposition of at least two spectra as is suggested by the change in shape of the spectrum during irradiation. Table 2 summarizes all positive and possible attributions for HPLC peaks detected after irradiation of an aqueous solution of CV (chromatograms in figure 3).

Journal of Physics: Conference Series **231** (2010) 012011 doi:10.1088/1742-6596/231/1/012011

Table 2: positive and possible attributions for the HPLC peaks detected after irradiation of an aqueous solution of CV.

Peak Number	RT/min	λ_{max}/nm	Positive Attribution	Possible Attribution
1	21.26	588	CV	
2	20.79	582	Mono-demethylated CV	
3	20.23	573	Bis-demethylated CV	
4	19.54	570	Tri-demethylated CV	
5	18.87	572	Tetra-demethylated CV	
6	18.05	562		Penta-demethylated CV
7	17.61	620		Diamond green-like compound or N-oxide derivative of CV
8	17.32	---	superimposition of two spectra	
9	16.89	597		Diamond green-like compound or N-oxide derivative of CV
10	16.34	554		Pararosaniline-like compound
11	15.48	579		Diamond green-like compound or N-oxide derivative of CV
12	14.29	543	pararosaniline	

3.2. CV on paper

3.2.1 *HPLC-PDA analysis.* Irradiation caused an evident fading of the colour of dyed paper. A quantification for the colour change, in terms of colorimetric L*a*b* values, is given later.

Chromatograms for CV extracted from paper before and after 340 hours of irradiation are shown in figure 7. Peak numbering is consistent with the one in figure 3. After 340 hours of irradiation a significant decrease in CV concentration was observed and other peaks had increased in intensity, showing that degradation products were being formed.

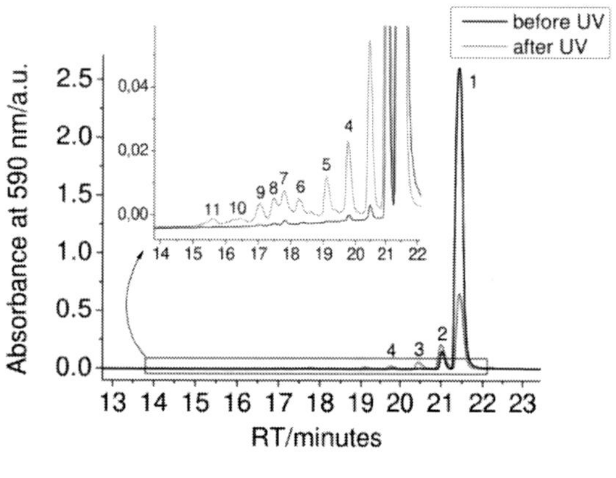

Figure 7: Chromatograms of CV extracted from paper before and after 340 hours of irradiation.

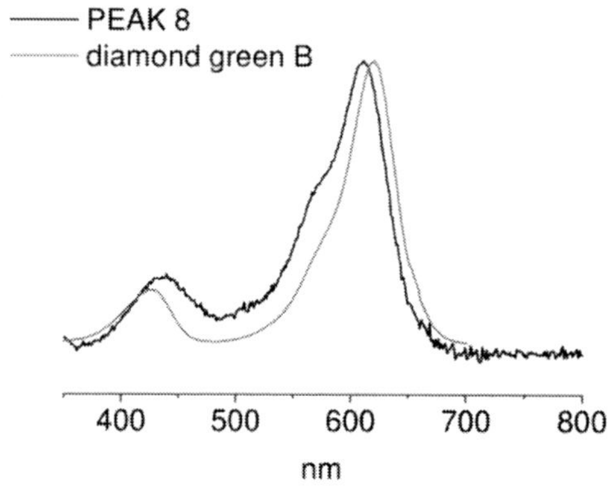

Figure 8: Normalized absorption spectra for peak 8 and diamond green B in the chromatogram of CV extracted from paper after 340 hours of irradiation.

The chromatogram of the extract obtained after irradiation of CV on paper was compared to that obtained after irradiation of CV in aqueous solution.

Except for the peaks having retention times between 14 and 15 minutes, the same peaks were obtained whether CV was irradiated in solution or applied to paper, indicating that the same degradation products are formed. With the only exception of peak 8, a comparison of absorption spectra for all the peaks confirms the analogy.

The absorption spectrum belonging to peak 8 (for CV on paper after irradiation) is similar to that of for diamond green B, but it has a shoulder. In figure 8 the absorption spectra belonging to peak 8 (RT=17.65 min) and diamond green (RT=19.76 min) are shown. In this case peak 8 is not due to the superposition of two different species as instead was the case in aqueous solution. Based on the retention times measured, the species responsible for peak 8 in figure 7 should be more polar than diamond green. However, due to the similarity of absorption spectra, they could be diamond green-like compounds: the higher polarity could be possibly due to a higher degree of demethylation of the species responsible for peak 8. The blue shift in the absorption maximum is consistent with this explanation.

To summarize, apart from peak 8, the same considerations made for CV in solution about peak 1 to 11 apply also for CV on paper (table 2).

Among the possible intermediates in the synthesis of CV and methyl violet there is Michler's ketone (figure 1D), which is of particular interest [8]. As a reagent in the synthesis of the dye, impurities of Michler's ketone could be present in the ink from the beginning. Moreover, Michler's ketone is believed to be a degradation product of CV and methyl violet and therefore it could also be formed on paper. Michler's ketone can accelerate CV degradation and for this reason the effect of this ketone on CV degradation has been investigated in our work. In the presence of Michler's ketone the consumption of CV during irradiation was higher. Moreover, MK is not a catalyst but is consumed during the photo-fading of CV. This is therefore a possible explanation for the absence of MK in the chromatograms for faded CV on model paper samples.

3.2.2. *FORS analysis.* A substantial discoloration was observed for CV on paper during irradiation as can be seen from the colorimetric L*a*b* values reported in table 3.

Table 3: L*, a*, b* values for CV on paper at different irradiation times.

	Before UV	16 h UV	33 h UV	79 h UV	102 h UV	126 h UV	149 h UV	176 h UV	200 h UV
L*	52,96	53,85	56,58	62,38	64,67	66,77	70,13	70,84	71,60
a*	21,08	19,99	13,53	8,98	6,55	6,09	4,50	3,10	3,03
b*	-62,10	-42,64	-37,98	-31,03	-29,08	-29,00	-25,65	-23,96	-23,30

The a* component decreased whereas b* and L* components increased upon irradiation giving a duller and lighter colour. Absorption spectra for CV on paper before and after irradiation are shown in figure 9. Upon irradiation both a decrease in absorption intensity and a change in spectral features were observed.

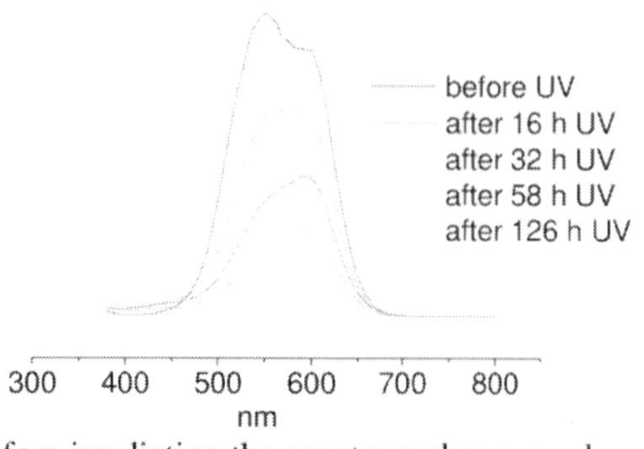

Figure 9: Absorption spectra for CV on paper before and after irradiation.

Before irradiation the spectrum shows an absorption maximum at around 550 nm and a pronounced shoulder at around 600 nm. Upon irradiation there is an increase of relative intensity of the band at 600 nm. This behaviour has already been observed by many authors for CV in solution [7,9,10] and it has been explained as an effect of the aggregation of dye molecules, the absorptions around 590 nm (α-band) and around 550 nm (β-band) being attributed to single monomers and dimers of CV respectively. At higher dye concentrations larger aggregates are formed, causing a shift in the absorption maximum toward shorter wavelengths. FORS therefore enabled both to measure the decrease in CV concentration on paper upon irradiation and to evaluate the aggregation type for the dye. The latter, besides influencing spectral features, plays a role also in determining the photochemical behaviour of dyes and for this reason the possibility of determining the aggregation type for CV on solid substrates with FORS renders this technique very interesting from a conservation point of view.

3.2.3. *HPLC analysis of 'Montmajour' and reliability of model samples.* Based on a previous HPLC analysis, the purple dye derived from the drawing 'Montmajour', made by Van Gogh (figure 2) consists of a mixture of methyl violet and fuchsine. Such a mixture can be explained either as the result of CV degradation on paper or as the result of the use of an ink already containing the mentioned dyes. To this regard, historically accurate reconstructions would be needed in order to verify these two hypotheses.

Finally, the result of HPLC analysis of a model sample of CV faded on paper was compared to the result of HPLC analysis of the sample coming from the drawing 'Montmajour' (figure 10). Interestingly, peak positions were very reproducible for the historical sample and the model sample (except for the RT range 22-25 minutes) giving evidence of the reliability of the model samples used.

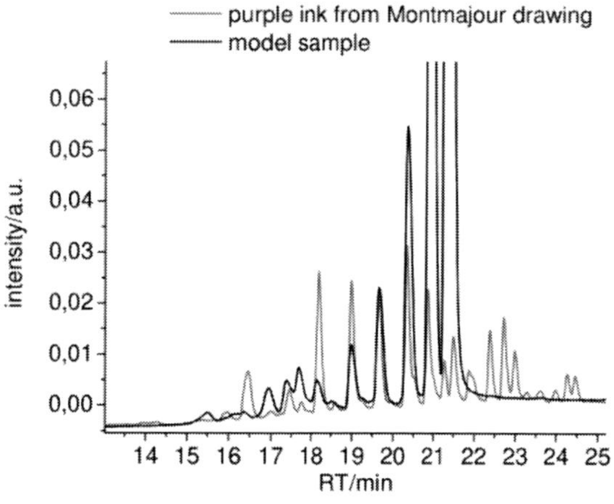

Figure 10: chromatograms for a model sample of CV faded on paper and for a sample of purple ink coming from the drawing 'Montmajour'.

4. Conclusion

The photo-induced degradation of CV was studied both in solution and on solid substrate via HPLC-PDA, LC-MS and FORS.

UV irradiation of CV in water caused a colour shift from purple to red. HPLC-PDA combined with LC-MS enabled to detect degradation products for CV in water. A series of demethylation products, from CV to pararosaniline (fully-demethylated CV) were detected. Moreover, diamond green-like spectra were observed which could be due to demethylation products of diamond green or N-oxides derivatives of CV.

With the applied experimental conditions, degradation products detected for CV on paper were the same as for CV in solution.

The photo-fading of CV in the presence of MK confirmed that the latter accelerates CV degradation and showed how MK is consumed during photo-fading.

Reflectance spectra enabled to express in term of colorimetric L*a*b* values the change in colour of CV on paper. Moreover, the changes in reflectance spectra during irradiation were attributed to changes in CV concentration and in CV aggregation type.

The comparison between the chromatogram for a model sample of CV faded on paper and for a sample of purple dye coming from 'Montmajour' drawing showed a very good consistency between results obtained from artificially aged model samples and those obtained from a naturally aged historical sample. Therefore, the method of ageing and analysis is reliable.

To conclude, the knowledge of the products formed during CV fading and their concentration ratios provides information about the original dye used for preparing the ink and consequently its original colour and its estimated date of production. It is however important to bear in mind that CV was often used in a mixture with various demethylated derivatives and it would be necessary to know the historical ink recipes and the production method of the dye in order not to misinterpret data on composition of historical ink samples.

Acknowledgements

The authors wish to thank Lingbeek Papierrestauratie Haarlem, for providing photo of figure 2. D. C. is grateful to Fondazione Aldo Gini for support.

References

[1] Havlinova B, Jancovicova V, Brezova V, Ceppan M and Turanova M 2006 Study of aging of arylmethane dyes by UV-Vis spectroscopy *Restaurator* **27** 24-34

[2] Allen N S 1994 Photofading and light stability of dyed and pigmented polymers *Polym. Degrad. Stab.* **44** 357-74

[3] H. Neevel, M. van Bommel, *Non-invasive analysis of Van Gogh's drawing inks*, ICOM committee for conservation, 15th triennial conference New Delhi, 2008, preprints volume I

[4] Gordon P F and Gregory P 1983 *Organic chemistry in colour* (Berlin: Springer-Verlag)

[5] Abbott L C, MacFaul P, Jansen L, Oakes J, Smith J R L and Moore J N 2001 Spectroscopic and photochemical studies of xanthene and azo dyes on surfaces: cellophane as a mimic of paper and cotton *Dyes pigments* **48** 49-56

[6] Nakamura R and Hida M 1982 Photoreaction of crystal violet in solution *Sen-I Gakkaishi* **38**. 183-90

[7] Duxbury D F 1993 The photochemistry and photophysics of triphenylmethane dyes in solid and liquid media *Chemical Reviews* **93** 381-433.

[8] Zollinger H 2003 *Color Chemistry, Syntheses, Properties, and Applications of Organic Dyes and Pigment* (Zurich: Wiley-VCH).

[9] Allen N S and McKeller J F 1980 *Photochemistry of dyed and pigmented polymers* (London: Applied Science Publishers)

[10] Stork W H J, Lippits G J M and Mandel M 1972 Association of crystal violet in aqueous solutions *J. Phys. Chem.* **76** 1772-75

Journal of Physics: Conference Series **231** (2010) 012012 doi:10.1088/1742-6596/231/1/012012

Investigation into Digital Print Hard Copy Quality, Longevity and Durability

Algy Kazlauciunas [1], Mingyu Zhang [1] and Jeff Ball [2]

[1] Department of Colour Science, School of Chemistry, University of Leeds LS2 9JT

[2] Nazdar Limited, Heaton Mersey, Stockport, Cheshire SK4 3EG

ccdak@leeds.ac.uk

Abstract. The main purpose of the research project was to investigate whether deskjet ink jet printing could produce A-4 size images on a par with conventional silver halide continuous tone photographic media prints. To this end, equivalent images were produced using both dye-based and pigment-based deskjet ink jet prints. To assist the research project, a multi-cultured array of observers were enlisted to give individual responses to a range of specific questions appertaining to the image characteristics of the A-4 size images produced. In addition, in order to assess the image stability and print durability of both the conventional silver halide prints and the ink jet prints, a range of laboratory tests were undertaken in the areas of light fastness, water fastness and both dry and wet rub fastness.

1. Introduction

Digital cameras have now taken over from conventional film cameras as the dominant form of image capture, whether that be via a phone camera, compact camera or single lens reflex camera. This has changed the landscape for the production of hard copy prints of these digitally captured images. One commonly used method of achieving hard copy prints is by means of a Minilab printer system such as the Fujifilm 700 series, which ranges from the compact LP 7000 model through to the high performance LP 7900 model. These printers use a combination of a high quality scanner, digital imaging processing software and RGB lasers, to create images on laser optimised silver halide papers such as the Fujicolor Crystal Archive brand.[1] The prints are commonly produced dry to dry in approximately four minutes, following development in a modified RA4 chemistry using CP-48S chemicals. However, many amateur, semi-professional, and professional photographers favour producing their own colour prints using either drop-on-demand technology deskjet ink jet printers [2, 3, 4, 5] or thermal dye transfer printers [6, 7], with ink jet printing being the more common technology in the United Kingdom.

2. Image Generation and Printing Procedure

Three individual images were recorded for the project. These comprised of a landscape photograph of Knaresborough viaduct (see figure 1); a group portrait image of six international students showing a range of different skin tones (see figure 2); and a multi-coloured image of predominantly fruit and vegetables (see figure 3). All three images were captured as RAW files using a Nikon D700 digital single lens reflex camera, fitted with a Sigma 18-50mm zoom lens. The first two images were captured in daylight and the third captured using tungsten photographic lamps (colour temperature 3200K).

Following post production in Photoshop CS4 and conversion to TIFF file format images, the images were printed as follows:

(a) A Fujifilm Frontier 700 series minilab printer was used to generate the silver halide prints on Fujifilm Crystal Archive glossy paper.

(b) An Epson Stylus Photo 1400 printer, fitted with an OEM ink set, was used to generate the dye-based ink jet prints on Epson Premium Glossy Photo paper.
(c) An Epson Stylus Photo R800 printer, fitted with an OEM ink set, was used to generate the pigment-based ink jet prints on Epson Premium Glossy Photo paper.

For the ink jet images, an X-Rite/Pantone ColorMunki photo spectrophotometer was utilised to calibrate both the PC monitor and for creating the profiles for the printer paper/ink combinations. [8, 9]

Figure 1. Landscape image

Figure 2. Group portrait image

Figure 3. Multi-coloured fruit/vegetable image

3. Observer Exercise

A total of thirty people of different nationalities were recruited for the observer part of the research project. The split in nationalities was ten European, ten Chinese and ten 'other nationalities' (participants from Iran, Pakistan, Korea and Saudi Arabia). The gender distribution was equal i.e. fifteen males and fifteen females.

Observation of the prints was undertaken using a standard neutral grey painted viewing cabinet, fitted with D65 daylight equivalent lighting (see figure 4).

Figure 4. Print viewing conditions

In order to compare the images generated by the different printing methods, a number of questions were created for each individual print series, and also for the combined images.

The questions generated for the observer exercise were as follows:

(a) Landscape image
Question 1. Which of the images is the most true to life?
Question 2. Which of the images do you prefer?
Question 3. Which of the images do you think has the best resolution?

(b) Group portrait image
Question 1. Which of the images do you think represents the correct skin tone?
Question 2. Which of the images is the most true to life, with respect to the overall colour?
Question 3. Which of the images do you prefer?
Question 4. Which of images do you think has the best resolution?

(c) Multi-coloured fruit/vegetable image
Question 1. Which of the images is the most true to life, with respect to the colour?
Question 2. Which of the images do you prefer?
Question 3. Which of the images do you think has the best resolution?

(d) Combined images
Question 1. Which of the six images do you think has the best resolution?
Question 2. Which of the six images do you think has the best contrast?
Question 3. Three of the six images are silver halide photographic images – please choose which three.

4. Laboratory Testing

In order to test the light fastness and durability of the individual prints a range of laboratory tests were undertaken, namely light fastness, water fastness, and both dry and wet rub fastness.

4.1 Light Fastness Testing

The light fastness testing was undertaken, as an accelerated ageing method, for assessing identical areas of the individual images, following exposure to artificial light.

The method undertaken involved the use of the Atlas Xenotest Alpha LM, fitted with a Xenon arc lamp. This test method is a British Standard / ISO method [10] developed primarily for the textile industry, yet was utilised here due to the fact that although there is a specified ISO standard for photographic prints [11], there is still no specific ISO standard for predicting the life of digital prints exposed to light on display, stored in albums or other dark places, or exposed to ambient ozone in homes or offices.[12] Thus, as with a number of research projects undertaken in the Department of Colour Science / Department of Colour and Polymer Chemistry at Leeds University [13, 14, 15, 16], this method of light fastness testing was employed.

Identical area sections were cut from the six individual prints (approximate size 4.5 x 1.5cm) and these were attached to standard white card (size 4.5 x 13cm). About 50% of each area section was then masked over by attaching a piece of white card along the length of the assembly. The card assembly was then placed into a metal sample holder which was placed into the Atlas Xenotest light fastness chamber.

In addition, a series of Blue Wool Standards was also prepared in identical fashion, to be exposed alongside the test samples. As for the test samples, the Blue Wool Standards were also cut to the same size (4.5 x 1.5cm) and attached to a piece of white card (size 4.5 x 13cm). The dyes utilised for the Blue Wool Standards 1 to 7 are tabulated in table 1 below.

Journal of Physics: Conference Series **231** (2010) 012012 doi:10.1088/1742-6596/231/1/012012

Table 1. Dyes for Blue Wool Standards 1 to 7

Blue Wool Standard	Dye Colorants – Colour Index Designation
1	C.I Acid Blue 104
2	C.I. Acid Blue 109
3	C.I. Acid Blue 83
4	C.I. Acid blue 121
5	C.I. Acid Blue 47
6	C.I. Acid Blue 23
7	C.I. Solubilised Vat Blue 5

The Xenon arc lamp, fitted in the Atlas Xenotest Alpha LM light fastness equipment, gives a very similar spectral distribution to natural daylight. The lamp itself is surrounded by seven infrared borosilicate filters. The following equipment conditions were used for the light fastness testing, with the test and reference samples being monitored every four hours until a Blue Wool Standard grade six had been reached.

- Equipment running mode – non-turning
- Chamber temperature - 20°C
- Relative humidity – 60%
- Input energy – 55

4.2 Water Fastness Testing

The water fastness testing procedure was used to evaluate the relative water resistance of the respective silver halide-based, dye-based and pigment-based images. The method used for this had been devised during a previous research project [17] and involved removal of specific areas from each of the printed images, followed by an equal volume of distilled water being placed onto the surface of the image via a pipette. Duration times of 30 seconds, 1 minute, 5 minutes and 15 minutes were used for the testing. A stop-watch was used to ensure time accuracy, and filter paper was used to soak-up the deposited water, following the specified time duration. The filter papers were then dried and retained for observation to see if any colorant had transferred from the surface-wetted print onto the filter paper. All samples were subsequently photographed. Any indication of colour transfer onto the filter paper would prove to be an indication of colorant migration and therefore reduced water resistance.

4.3 Dry Rub Fastness Testing

The dry rub fastness test procedure mimicked that procedure developed for a previous research project [17] and was used to evaluate the relative mar resistance of the individual prints under dry conditions.

The first stage was to cut-out individual pieces from the same area of each of the prints for direct comparison. The samples were fixed in a horizontal position and a single cotton bud used for the rub fastness procedure. This involved rubbing the surface one hundred times in a circular manner, ensuring that the strength applied was uniform for all the samples. All test samples were retained for visual comparison and a series of photomicrographs of the rub tested print surface areas were recorded.

4.4 Wet Rub Fastness Testing

The wet rub fastness procedure was also developed for a previous research project [17] and was used to evaluate the relative mar resistance of the prints under wet conditions.

As for the dry rub fastness procedure, individual pieces were cut from the same area of the prints for direct comparison, and the procedure was identical with the exception that the cotton buds were thoroughly wetted in distilled water prior to undertaking the one hundred circular motions. Again, all samples were retained for visual comparison and a series of photomicrographs recorded of the rub-tested print surface areas were recorded.

5. Experimental Results

5.1 Observer Exercise

Table 2 below shows the distribution of the observer response to the questions relating to the landscape prints.

Table 2. Image observation response for the landscape prints

Observer nationality	Question number	Ink Jet Dye-based image	Ink Jet Pigment-based image	Silver halide-based image
European	1	4	3	3
	2	8	0	2
	3	7	0	3
Chinese	1	6	4	0
	2	9	0	1
	3	7	2	1
'Other nationalities'	1	6	1	3
	2	8	2	0
	3	9	0	1

Table 3 overleaf shows the distribution of the observer response to the questions relating to the group portrait prints.

Table 3. Image observation response for the group portraits

Observer nationality	Question number	Ink Jet Dye-based image	Ink Jet Pigment-based image	Silver halide-based image
European	1	4	1	5
	2	5	3	2
	3	7	0	3
	4	8	2	0
Chinese	1	3	2	5
	2	5	3	2
	3	6	2	2
	4	6	4	0
'Other nationalities'	1	2	1	7
	2	3	3	4
	3	7	1	2
	4	6	3	1

Table 4 below shows the distribution of the observer response to the questions relating to the multi-coloured fruit/vegetable prints.

Table 4. Image observation response for the multi-coloured images

Observer nationality	Question number	Ink Jet Dye-based image	Ink Jet Pigment-based image	Silver halide-based image
European	1	6	3	1
	2	8	2	0
	3	9	0	1
Chinese	1	5	4	1
	2	10	0	0
	3	9	1	0
'Other nationalities'	1	6	3	1
	2	8	2	0
	3	7	3	0

Table 5 overleaf shows the observation response to the questions regarding the combined images.

Table 5. Image observation response to the combined images

Observer nationality	Question number	L – IJ -DB	L - IJ - PB	L- SHB	P-IJ -DB	P-IJ -PB	P- SHB	MC-IJ -DB	MC-IJ -PB	MC SHB
European	1	7	0	0	1	0	0	2	0	0
	2	1	0	0	4	0	0	4	0	1
	3	3	1	6	2	2	6	3	3	4
Chinese	1	1	0	0	4	1	0	4	0	0
	2	1	0	0	5	0	0	4	0	0
	3	3	4	3	6	1	3	4	3	3
'Other nationality'	1	1	0	0	3	1	1	4	0	0
	2	2	0	0	4	0	1	3	0	0
	3	7	0	3	6	1	3	7	1	2

Key for table 5:
L-IJ – DB = Landscape/Ink Jet/Dye based.
L-IJ – PB = Landscape/Ink Jet/Pigment based.
L-SHB = Landscape/Silver halide based.
P-IJ – DB = Group Portrait/Ink Jet/Dye based.
P-IJ – PB = Group Portrait/Ink Jet/Pigment based.
P-SHB = Group Portrait/Silver halide based.
MC-IJ – DB = Multi-Coloured/Ink Jet/Dye based.
MC-IJ – PB = Multi-Coloured/Ink Jet/Pigment based.
MC-SHB = Multi-Coloured/Silver halide based.

Table 6 below illustrates the breakdown of table 5 into the basic selections of either ink jet dye-based, ink jet pigment based, or silver halide based.

Table 6. Basic print selections

Observer nationality	Question number	Ink Jet Dye-based image	Ink Jet Pigment-based image	Silver halide-based image
European	1	10	0	0
	2	9	0	1
	3	8	6	16
Chinese	1	9	1	0
	2	10	0	0
	3	13	8	9
'Other nationality'	1	8	1	1
	2	9	0	1
	3	20	2	8

5.2 Laboratory Testing Results

5.2.1 Light Fastness Testing. Table 7 overleaf illustrates the light fastness testing results obtained for the print samples.

Table 7. Light fastness test results

Sample reference	Blue Wool Standard Grade
Landscape – Ink Jet Dye-based	6
Landscape – Ink Jet Pigment-based	6
Landscape - Silver halide-based	6
Group portrait – Ink Jet Dye-based	6
Group portrait – Ink Jet Pigment-based	6
Group portrait – Silver halide based	6
Multi-coloured – Ink Jet Dye-based	6
Multi-coloured – Ink Jet Pigment-based	6
Multi-coloured – Silver halide-based	6

5.2.2 Water Fastness Testing. There was no evidence of the transfer of the colorant onto the filter paper, over the full time duration, for either the ink jet pigment-based prints or the silver halide-based prints. However, although there was no evidence for the transfer of colorant from the ink jet dye-based prints over the time duration of 30 seconds and 1 minute, there was evidence of a minor quantity of colorant having transferred to the filter paper after 5 minutes and 15 minutes.

5.2.3 Dry Rub Fastness Testing. Following the dry rub fastness process for the ink jet dye-based print, the tip of the cotton bud showed no evidence for any of the colorant having transferred across from the print. However, for both the ink jet pigment-based print and the silver halide-based print, some black colorant was transferred onto the respective cotton bud tips used for the test. Inspection of the respective prints, following the rub fastness process, did not show any undue change to their respective surfaces.

5.2.4 Wet Rub Fastness Testing. Following the wet rub fastness process on the ink jet pigment-based print, there was no evidence of pigment having transferred across to the cotton bud tip. Thus the wet mar resistance properties of the pigment-based print are better than the dry mar resistance.

For the ink jet dye-based print, the colorant proved relatively easy to rub off and therefore, the wet cotton bud tip darkened quickly.

The silver halide print proved to have the weakest wet mar resistance as most of the colorant transferred across onto the cotton bud tip, following the 100 cycle rubbing procedure. This meant that the rubbed area of the print was practically clear of the presence of colorant, revealing the surface of the substrate.

6. Discussion and Conclusions

6.1 Observer Exercise

For the observation exercise, comparing the A-4 size images produced on the Epson ink jet dye-based printer, the Epson ink jet pigment-based printer and the Fujifilm Minilab digital system, the following conclusions can be drawn:

(a) With respect to the prints of the landscape image, there is a very strong agreement between the three observer groups that the dye-based ink jet print is their preferred choice (80-90%) and also deemed to be of the highest resolution (70%). However, although this particular print was also the preferred choice in the true-to-life category, the result was not as clear-cut. Whereas 60% of both the Chinese and the 'Other nationalities' chose this print, only 40% of the European observers did so. From the results it is clear that the true-to-life question appears to have been the most difficult to answer. Whereas none of the Chinese observers chose the silver halide print in this category, 30% of the observers for both the European and the 'Other nationalities' did so. However, overall across the board, the pigment-based ink jet print was marginally in front of the silver halide print for second choice.

For the group portrait image, once again there is very good agreement between the three observer groups with respect to their preferred print (60-70%) and the print deemed to be of the highest resolution (60-80%). The other two questions however received a more varied response. Although the highest percentage of the observers chose the silver halide print as their choice for best skin tone representation, the result was not unanimous, with overall 47% of the thirty observers choosing the silver halide print, 33% the dye-based ink jet print and 30% the pigment-based ink jet print. The true-to-life, with respect to overall colour, choice of print also proved to be a difficult decision to make for the observers. For this category, the dye-based ink jet print was the favoured choice for both the European and Chinese observers (both 50%). However, the silver halide print was the favoured choice for the 'Other nationality' observers (40%). Overall, 43% chose the dye-based ink jet print, 30% chose the pigment-based print and 27% chose the silver halide print.

For the multi-coloured fruit/vegetable image, the response from each of the observer groups was very similar for each of the questions presented. The dye-based ink jet print was the preferred choice in each of the categories i.e. highest resolution print (70-90%), preferred image (80-100%) and true-to-life image (50-60%). Overall, 57% of the observers chose the dye-based ink jet print, 33% chose the pigment-based ink jet print and only 10% preferred the silver halide print.

When the six prints were placed together for viewing, the previous trend continued, with 80-100% of all observers choosing a dye-based ink jet print as having the highest resolution. In addition, between 90-100% of the observers chose a dye-based ink jet print in the highest contrast category. However, with respect to being able to pick out the three silver halide prints from the ink jet prints, only the European observers had a modicum of success with a correct response rate of 57%. Both the Chinese and the 'Other nationalities' observer groups had a meagre success rate of only 27%.

The overall verdict from the observers when combining the data from all the images is illustrated below in table 8.

Table 8. Observer data from combined images

Question category	Dye-based ink jet print (%)	Pigment-based ink jet print (%)	Silver halide-based print (%)
True-to-life image	51	30	19
Preferred image	79	10	11
Highest resolution image	79	14	7
Best skin tone representation	30	13	57
Highest contrast image	93	0	7

6.2 Laboratory Testing

The laboratory testing undertaken on the ink jet dye-based, the ink jet pigment-based and the silver-halide based prints, revealed the following findings:

(a) The light fastness testing results indicate that all the different print types achieved at least a Blue Wool standard value of 6 (see table 7). The results achieved are not surprising for an ink jet pigment-based print and a silver halide based print, but are a remarkable testament to the light stability of the Clariant dyes used in the Claria Epson OEM dye-based inks. Two web sites however indicate that prints made with Claria inks will last for 98 years in a photo frame and 200 years in a photo album. [18, 19] Part of the reason for the high light fastness values associated with the black, magenta and cyan dyes appears to be the increased molecular weight of the respective structures. [20, 21, 22]

(b) For the water fastness testing exercise, there was no transfer of colorant onto the filter paper from either the ink jet pigment-based print or the silver halide-based print. However, there was transfer of colorant onto the filter paper, following exposure for 5 and 15 minutes, for the ink jet dye-based print. The levels transferred though, were of a very minor quantity and did not appear to affect the surface of the print.

(c) With respect to the dry rub fastness testing, the ink jet dye-based print showed no evidence for the transfer of colorant to the cotton bud tip, whereas both the ink jet pigment-based print and the silver halide-based print did show clear evidence of colorant removal.

(d) For the wet rub fastness testing, the ink jet pigment-based print proved to be the most successful with no colorant having been transferred. Whereas, both the ink jet dye-based print and the silver halide-based print showed transfer of colorant onto the cotton bud tip. The ink jet dye-based print showed little or no surface damage, when viewed under an optical microscope (see figure 5), whereas the silver halide-based showed clear evidence for surface print damage when viewed under identical conditions (see figure 6).

(e) Following the combined observer and laboratory testing exercises, the digital printing method that takes most of the honours is the ink jet dye-based system. The only negative aspect with respect to the laboratory testing exercise was a minor weakness in the water fastness test, and some transfer of colorant in the wet rub fastness exercise. However, in the observer exercise this printing method was the overwhelming favourite in all categories, with the exception of skin tone representation. In this category the silver halide-based print was the favoured choice. Thus, despite the improvement in skin tone representation for ink jet dye-based prints following the introduction of light magenta and light cyan inks, it would seem that this has not put the system on a par with silver halide technology. This certainly illustrates why the current image permanence test methods in photography, such as ISO 18909:2006, really require full tonal scale human skin tone colours in the fading evaluation, especially when approximately 80% of all amateur photographs include people in the scene and for all professional wedding and portrait photographers, this increases to 100%. [23]

Figure 5. Photomicrograph of ink jet dye-based print surface following wet rub fastness testing

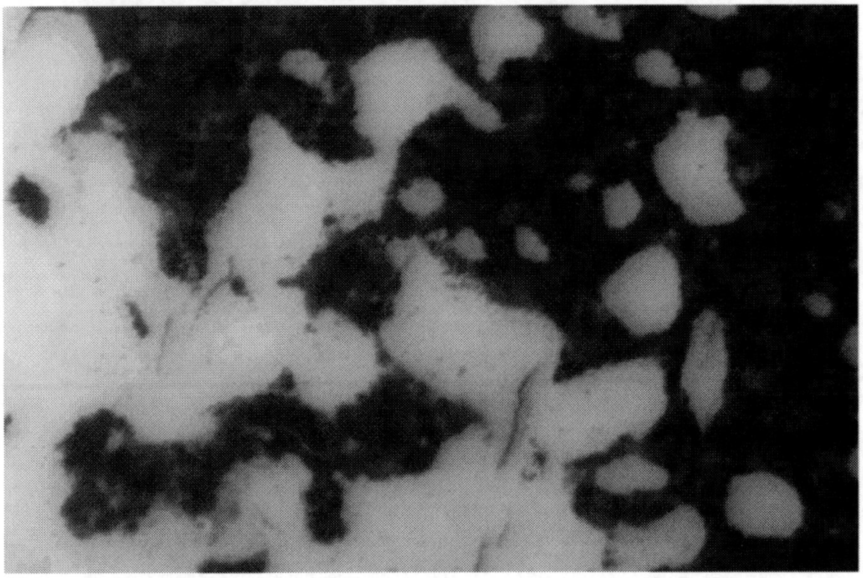

Figure 6. Photomicrograph of silver halide-based print surface following wet rub fastness testing

References

[1] Kimura T Doi A Kojima T Kubu M and Igarashi A 2006 *Color Desktop Printer Technology* ed N Ohta and M Rosen (Boca Raton: CRC/Taylor & Francis Group) pp 220-221.

[2] Kenyon R W 1996 *Chemistry and Technology of Printing and Imaging Systems* ed P Gregory (Glasgow: Blackie Academic & Professional) pp 113-118.

[3] Kipphan H ed 2001 *Handbook of Print Media* (New York: Springer-Verlag) pp 715-720.

[4] Wnek W J 2002 *Handbook of Imaging Materials* 2nd Edition ed A S Diamond and D S Weiss (New York: Marcel Dekker) pp 457-463.

[5] Johnson H 2003 *Mastering Digital Printing* (Ohio: Muska & Lipman) pp 192-195.

[6] Bamfield P and Hutchings M G 2010 *Chromic Phenomena – Technological Applications of Colour Chemistry* 2nd Edition (Cambridge: RSC Publishing) pp 224-226.

[7] Bradbury R 1996 *Chemistry and Technology of Printing and Imaging Systems* ed P Gregory (Glasgow: Blackie Academic & Professional) pp 151-163.

[8] Sharma A 2004 *Understanding Color Management* (New York: Thompson/Delmar Learning) Chapters 7 & 8.

[9] Rodney A 2005 *Color Management for Photographers* (Oxford: Focal Press) Chapters 3 & 6.

[10] BS EN ISO 105-BO2: 1999 *Colour fastness to artificial light: Xenon arc fading lamp test.*

[11] ISO 18909: 2006 *Photograph – Processed photographic films and paper prints – methods for measuring image stability.*

[12] Wilhelm H Armah K Shklyarov D and Stahl B 2009 Improved Test Methods for Evaluating the Permanence of Digitally-Printed Photographs *Proc. Imag. Conf. Japan*, p 213.

[13] Park J R 2000 *The* Archival Quality of Ink Jet Printed Images *BSc Research Dissertation, Department of Colour & Polymer Chemistry, University of Leeds.*

[14] Aspinall C P 2005 Ink/Ink Receivable Layer Performance of Dye & Pigment Based Photo-realistic Ink Jet Printed Images *BSc Research Dissertation, Department of Colour & Polymer Chemistry, Leeds University.*

[15] Wei Z 2007 Incorporation of UV Absorbers within Ink Jet Ink Formulations and Ink Jet Ink Receivable Layers *MSc Research Dissertation, Department of Colour Science, Leeds University.*

[16] Li Z 2008 Analysis and Characterisation of Ink Jet Dye Ink Formulations, together with the Longevity Performance of various Dye Ink/Substrate Combinations *BSc Research Dissertation, Department of Colour Science, Leeds University.*

[17] Sheng Y 2008 A Carbon Footprint Study for the Production and Recycling of Billboard Poster Materials *MSc Research Dissertation, Department of Colour Science, Leeds University.*

[18] www.epson.co.uk/Explore-Epson/1223288350198.

[19] www.wilhelm-research.com

[20] US Pat. 2007/0263056 A1.

[21] US Pat. 7,279,034 B2.

[22] US Pat. 2007/0263055 A1.

[23] Wilhelm H and Shklyarov D 2007 Evaluating the Image Permanence of Full Tonal Scale Human Skintone Colors in Photographs using the CIELAB Colorimetry Based WIR i-Star Retained Image Appearance Metric *Proc. Int. Conf. NIP 23rdDigital Printing Technologies*, pp 743-748.

Journal of Physics: Conference Series **231** (2010) 012013 doi:10.1088/1742-6596/231/1/012013

The Aesthetics of the Negative or; The Negative Dialectics of Photography

Ana Peraica

Makarska 1, 21 000 Split

aperaica@ffri.hr

Abstract

This essay gives technical details on two photographic experiments, Positiveland (2009) and Negativeland (2009), intended to indicate physical constraints of photographic aesthetics, in terms of material, style and today's condition. Namely, what is commonly seen as the aesthetics of the vintage photograph may be a non-original, romanticised condition of the very photograph splitting apart. With different techniques, in recording and digital post-processing elementary slices of images can be analysed and compared, thus allowing entering deeply into the author's signature and original choices. This solution may be important in conservation and preservation of historic image, but as well in interpretation of photographs.

1. Introduction

Photographic theory and criticism of the early XX century may appear vague if the original set of reference, being the technology of the era [1], is not taken into account. This appears the case when referring to the concept of "automatic camera," as seen by early photography theorists. Namely, Stieglitz [2] and others, in criticising the fashion of snapshot photography as installing the "era of Button-pressers," were referring to a Brownie camera [2]. These cameras were not, as did compact ones in the eighties, having an inbuilt automatic light-meter nor did they have the auto-focus option that would simplify the use. Still, one thing they share in common with contemporary versions is - both used a rolled film, which simplified the production of photographs [1]. Kodak's invention reduced all previously demanded special knowledge for the preparation of negatives. This specific automatism changed furthermore with digital imaging development. The automatic camera of the eighties, seen from the contemporary digital age, seems more complicated, in terms of film handling, no matter how easier was to develop the roll of a film. With "digital shift" the negative or the photographic prototype, disappeared. Or, the world of images has lost own prototypes [3].

Similarly to the notion of automatic camera, interpretation may find problems with definition of photographic negatives and positives, for there is a variety of materials and chemical processes used during the last hundred and fifty years since the invention of photography. The world of negatives still, was rarely depicted theoretically. Only few authors have referred to the negatives [4] out of their technical necessity inside the production of the photographic image, all until recently. In some latter discussions, precisely condensing around the invention of the autofocus compact camera, the split induced by the division of labour on "button pressers" and "laboratory workers" was obvious. Walton [5] has seen the dark chamber as a place of conspiracy, similar to the one of Searle's Chinese room [6]. Needless to say the dark chamber, seen as the very realisation of the Plato's cave, was taken as the paradigm of ideology. Both [7] and Marx have suggested the relationship [8], one analysing conscience, while the other ideology. Still in terms of developing the theory of the negative unavoidable theorists are, aside Hegel, Adorno [9] Zizek [10] and Nancy [11].

In the upcoming elaborate I would focus on the material level of the negative – positive dialectics, the one present in photography. I will focus on comparison of original negatives and vintage prints, their conservation and digitalisation, by using digital technologies, thus framing the field of theoretical discussion and giving advices for practical handling of negatives and positives.

My meta-discourse would be photography itself, or the digital equipment available in restoration and conservation of photographs. With paper positives, or photographs, digitalisation seems simple, in terms of having a set of simple instructions suggested by professionals in conservation, but results of scanning positives are hardly near results of the research into original negatives. Namely, what can appear as the style of the romanticised vintage photograph may as well be the signal of its decay, while the original image is inaccessible.

To be able to show this thesis I've recorded various conditions of paper positive as well as negative behaviour in photographing. These two sets I've named Positiveland and Negativeland.

1.1. Rephotographing

Both the Positiveland and the Negativeland, are produced by rephotographing. The concept of rephotographing itself was in Postmodern referred to two different processes; rephotographing the scene represented by the photograph and rephotographing the paper representation or - the photograph itself. Rephotographing of the first kind is often used in archival and documentation purposes where the landscape photographs are repeated in order to record the very change of the landscape after the urbanisation processes.§ The other type of rephotographing, one of the paper-positive, was quite Postmodern in the specific repetition with a change and a question of authors rights. There were orientated to the re-emerging theory of photography with a subsequent theory of the author, as appearing, after Benjamin, in theories of authors [12], as written by Barthes [13], Foucault [14] and Derrida [15].

The difference of the two is clear, paradoxically enough, in terms of violation of rights; one violates the very author's idea or right on the scene, while the other – the copyright of the work. In both types of photographing the authorship may lead to paradoxes. ** Crimp [16] retells: "Levine has said that, when she showed her photographs to a friend, he remarked that they only made him want to see the originals. "Of course," she replied, "and the originals make you want to see that little boy, but when you see the boy, the art is gone". Still, as also Levine's work has shown; interpretations can happen even in the accurate 1:1 re-photographing on the level of conceptual choice of the second photographer as author decision. Krauss cited Barthes in referring to Levine's work "To depict is to ... refer not from a language to a referent, but from one code to another. Thus realism consists not in copying the real but in copying a (depicted) copy .. Through secondary mimesis [realism] copies what is already a copy."[17]††

Positiveland and Negativeland are made in the second term of rephotographing, the one having own interpretation of the material. The main task of both cycles was to answer a set of questions deriving from the difference of scanning and re-recording photographs, as, the depth of the image, the photographic material itself, original interventions into it etc...

2. Re-Recording

The difference between scanning photographs and re-photographing is vast, I will show by a couple of re-photographing tests. Strong light of scanning lenses can pass through the old paper positive and record traces of the background on the foreground within the same image. An emitting pink replaces the natural pale brown colour of the old photo paper. Paper image rather reflects emitting pink as if the paper was overexposed to the light. Intuitively, seeing pinkish surround, one cannot but conclude the photograph is damaged by the very process of scanning with the strength of light and its temperature. It literally looks as if it is burned.

The neglecting of the original material in interpretation may lead to many mistakes, especially when these techniques are important to the very aesthetics of the image, as for example differences of gelatine silver, albumen, collodium process, or other. These final medium of the image is especially important in early aesthetics of photography [18].

Furthermore, with digital correcting of scanned images what is brought in front is only the image whose original condition can hardly be depicted, but relies more on the aesthetic presumption of the re-photographer. The difference between the paper and the original photographer's will comes more visible when comparing positives and negatives. The original image, as recorded on the negative is smooth, while the very destroyed paper positive shows up a large interest in geometric shapes of deep shadows, high contrasts and a contrasting retouch in small details. This Modernist aesthetics can hardly be programmed in the automatic digital processing correction, as with auto-contrast, as the image itself is reworked within the process of developing but also in additional work.

2.1. Rephotographing

By rephotographing the damage of the old print is avoided. The contact is not direct. Still some damage may appear with handling and lighting.

2.1.1. Lenses. Chosen lenses of camera are giving more details than any process of scanning can. In Positiveland, when focusing on material, I have used manual telephoto lenses to be able to concentrate on different details of the paper image, such as material, depth of the image, but also different conditions they are found in, as being damaged by scars, holes, paper disintegration etc... In Negativeland, to the contrary, I have used wide view lens which can give more of general sharpness in the near field or the first plan. Both lenses used were fixed. The full connection of background and foreground can be recorded using stronger photographic lenses.

2.1.2. Light. Positiveland in recorded in a horizontal lightening, as in ordinary view, allowing layers or slices of photographic image to be visible at different point. This was possible only with narrowing the depth of the field, using shutter speed larger than 2000 and large exposure, optionally. Negativeland is recorded against the direct sunlight, in counter-light, as only then they are visible. Furthermore they are over-exposed, using the method of long exposure to gain the light into the dark glass, so the surround appears bleached, but once inverted – the surround will become black.

2.1.3. Post-Processing. While Positiveland is not post-processed, Negativeland is tied to additional processing. In Negativeland large-size negatives are rephotographed and inverted back into positives, but with its near surround, drawing the world into the negative. As images in the Negativeland and Positiveland would appear very similar, I have further used differences in lighting, which have made Negativeland appearing in definitions of cold colours, while Positiveland, recorded in the smooth daylight appearing, in warmer scale of colours. This colour specificity is reached by the paradox inbuilt in the medium.

2.1.3.1. Post-Processing Errors. Aside inversion of the negative, all colour corrections in Negativeland are put down so the black which is control colour can define the image on the negative, making at the same time all the surrounds hysteric in colours. Also all the contrasts are put down to get more of nuances of fragile black and white image. So, either a glass negative or the surrounding world is recognisable and visible. The negotiation between the two was impossible.

As time since the original negative has changed, as well as colour codes of each medium, the system error of transcription occurs on most of them. Rerecording negatives was indeed used in some processes to stabilise the medium, but usually making the same copy.‡‡ Rarely it is for manipulating contrast or exposure between black and white and dia-films, where again the colour function is narrowed back to the black and white.§§ The difference of film technologies and digital ones is also in the colour definition which prevents such an easy transfer, except for the black and white. Namely, CMYK technology recognised black as key colour, while not recognising accurately all other colours of the RGB spectrum. This was the reason of the difference of the monochrome films once with methods for the film-colour processing. So, recording glass negatives, which are monochrome, in CMYK produces a distortion of the real-world changing its colours in the key colours of its spectrum, while the real world and colour films appear in the negative which never existed in the real world as CMYK was not used in the negative process, so the negatives appear visually more similar to the ultra-violet photography positive than the ordinary film negative.

3. Photographic Tests

3.1. Positiveland

People hold photographs in their wallets... They also draw on them or write. But they also kiss them or tear them apart. Anyhow, they make relationships - to them and no matter how a photograph is treated, it would age as people do. It is the world of time. Still, while in painting the patina is giving an additional value, in photographs it is treated as decay. Some of positives in the archive seemed much older than their original negatives. Some were splitting apart or having different sorts of mechanical damage as scars, scratches, holes, but also interventions. These conditions were recorded in order to compare positives and negatives.

A special focus has been given to the print. Under smooth daylight in counter-view, the silver layer that reflects into a scanner when these images are scanned, is emitting the oxidation, turning images black and showing precisely paper condition.

In most of cases the reason was a thinner paper used in atelier's hyper-production, all until the image is chosen to be printed on in large format. But then the glue on the back of them, attaching them to the album or the background in frame has done even worst. Different behaviour of paper and textile has conditioned the different depth of the image. Although paper and textile positives are having an un-transparent texture, not allowing the projection the negatives do, they still possess a certain transparency visible once orientated against the sun. In direct backlight it is possible to choose the layer to record; the image, the surface with the scar...

And the light again makes the inner image to the image, as with shadows of the negative which are images as well, but these are made, as the paper is not forgetting – only once. In small frames that staid in our window-shop for a decade I've found shadows of images of two children in separate frame. A boy and a girl imprinted on the red background by the solarisation process. They were the only "the positive of the positive." While the original images bleached the background carried the image for itself as the ghostly reminder of once a photograph and then again, as in Levine's story; asking for the original image. But, contrary to Levine's work, the original image here has, as well as probably boys and girls; disappeared. But, if I was asked I would ask for the negative rather than the original boy. But whatever asked the situation cannot be strange as with finding a negative without own picture, or finding an image that was never integrated into reality.

3.2. Negativeland

Negativeland appears as the world bringing up distant past in vivid lightening, which as the turn effect obscures the present layout in front of the camera. The hallucinating effect is stronger as negatives are transparent, so the light passes through them making shadows which are again images – the original photographs themselves at the first place. The transparency allows to see the image from both sides Making a number of projections they come multi-dimensional, as the same image in negative is visible from both sides, which makes two images in projection. So, effects may resemble combination printing, although they are not ones.***

4. Conclusion; Towards the Metaphysics of the Negative

The number of positives in the real world is bigger than the one of negatives kept. Most of them, even those whose subjects are forgotten are integrated into reality. But, negatives appear as the fallout of processes. Negatives, as "memories of memories" or long-term memories to the positives which have split apart of disappeared, were for a long time seen as just as they were black-holes. Still, some serious theoretical insight may be interesting for this conference, is found within the physics.

In the philosophical argument of time travel, which can be infinitively connected, an interesting debate on photographic negatives has been initiated decades ago. It has played a significant role in time-travel philosophy based on Wheeler and Feynman assumptions of continuous space.[19] In Maudlin's revision of the modal argument, a standard black and white film is shooting results arriving from the time machine. [20] The first arrival is a developed film, which is shot and developed again and then is sent back to the time machine. The positive of the positive is set to come out at the very precise moment when the shot of developed negative was done. Author claims; the result is the negative which is on positive.

This argument is extremely interesting for those who have the laboratory practice. The input in positive, which was turned to a positive showing negative is actually a common dia-positive with complicated process. Whatever the result is, the author claims – positive of negative is still dependant on the original in a physical way, i.e. it is continuous. The film preserves a certain amount of gray which is never "translated" into own opposition.††† And this was proven even more radical in the conversion of colours. But, for claiming "Photography can not be used to create time travel paradoxes either!" Maudin might not be correct in physical sense. One can find other metaphorical visual meanings into it [19]

Namely, experimenting with negatives of negatives even today resembles a concept of the "looped time." On the serial Negativeland suddenly the distant past was re-appearing from negatives alive as if it happens in present time.

And there the notion of hidden appears back again with famous Freud's recognition. In a certain sense Negativeland seems to be the world of photographic meta-physics. So the technical problem arrives if; the negative which is surpassed technology, draws implications on our physics, in terms of meta-physics. He has noticed the obvious split between the conscious and subconscious hunting the photography, saying: „In order to form a picture of this vicissitude, let us assume that every mental process... exists to begin with in an unconscious stage or phase and that it is only from there that the process passes over into the conscious phase, just as a photographic picture begins as a negative and only becomes a picture after being formed into a positive. Not every negative, however, necessarily becomes a positive; nor is it necessary that every unconscious mental process should turn into a conscious one. This may be advantageously expressed by saying that an individual process belongs to begin with to the system of the unconscious and can then, in certain circumstances, pass over into the system of the conscious.[21, 22]

Taking in mind this subconscious and conscious relationship, I would conclude; the research into negatives, especially as enforced by digital imaging possibilities, in terms of post-processing is important way of finding hidden and oppressed within the history of photography itself. Only by comparing negatives and positives some important facts on the material, its condition, but also re-working may be found out. This would appear the most important in cases of revisionism in photography [23]. By recovering data all post-photographic interventions, being they the ones made by author or the editor, can be précised.

References

[1] Ford C and Steinorth K 1988 *You Press The Button, We Do The Rest: The Birth of Snapshot Photography* (London: Dirk Nishen Publishing)

[2] Stieglitz A 1897 *The Hand Camera–Its Present Importance.* The American Annual of Photography pp 18-27

[3] Baudrillard J 2000 [cited 1. 5. 2009] *Photography, Or The Writing Of Light* Available from: http://www.ctheory.net/articles.aspx?id=126.

[4] Adams A and Baker R *The negative.* The New Ansel Adams photography series bk.2. 1981, (Boston Little, Brown) 1983. xiii, p 272

[5] Martin E 1986 *On Seeing Walton's Great-Grandfather.* Critical Inquiry **12**(4) pp 796-800.

[6] Searle J R 1983 *Intentionality: An essay in the philosophy of mind* (Cambridge: Cambridge University Press)

[7] Locke J and Pringle-Pattison A S 1978 *An essay concerning human understanding*. [Abridged] ed. / abridged and edited by A.S. Pringle-Pattison. (Hassocks: Harvester Press [etc.]) [1], xlviii, p 380

[8] Marx K and Elster J 1986 *Karl Marx: a reader* (Cambridge: Cambridge University Press) vi p 345
[9] Adorno T W *Negative dialectics* 1973 (London: Routledge) 1990. [432]p.
[10] Žižek S 1993 *Tarrying with the negative: Kant, Hegel, and the critique of ideology*. (Durham, NC: Duke University Press)
[11] Nancy J.-L 2002 *Hegel: the restlessness of the negative* (Minneapolis ; London: University of Minnesota Press) xxix p 124
[12] Burke J M 1989 *The death and return of the author: criticism and subjectivity in Barthes, Foucault and Derrida* (University of Edinburgh)
[13] Barthes R 1977 *The Death of the Author*, in *Image - Text - Music* ed R Barthes
[14] Foucault M 1977 *What is an Author*, in *Language, Counter Memory, Practice* ed M Foucault (Blackwell: Oxford)
[15] Derrida J 1995 *On the Name* ed T Dutoit. (Stanford: Stanford University Press)
[16] Crimp D 1980 *The Photographic Activity of Postmodernism* October, **15** pp 91-101.
[17] Krauss R.E 1984 *The originality of the avant-garde and other modernist myths* (Cambridge, Mass.: MIT Press)
[18] Rosen J 1987 *The Printed Photograph and the Logic of Progress in Nineteenth-Century France* Art Journal, **46**(4 - The Political Unconscious in Nineteenth-Century Art): pp 305-311.
[19] Kelly T 2006 [cited 2009 1.5.2009] *Standford Encyclopedia of Philosophy online* Available from: http://plato.stanford.edu/entries/evidence/.
[20] Maudlin T 2002 *Quantum non-locality and relativity: metaphysical intimations of modern physics*. 2nd ed. (Malden, Mass. Oxford: Blackwell) xi p 281
[21] Cadava E 1992 *Words of Light: Theses on the Photography of History*. Source: Diacritics, **22** (No. 3/4, Commemorating Walter Benjamin) pp 85-114
[22] Cadava E 1997 *The Words of Light: Theses on the Photography of History* (Princeton: Princeton University Press)
[23] King D 1997 *The Commissar Vanishes. The Falsification of Photographs and Art in Stalin's Russia.* (Metropolitan Books Henry Holt and Company)

§ Such were Rauchenberg's Re-photographing Atget project, Andrzej Maciejewski's, After Notman (2003), or various artists in the book by Jean Provencher Que´bec, les images te´moignent (2001), rephotographing the work of Jocelyn Paquet on Quebec, as well as Mark Klett's Rephotographic Survey Project (1977 and 1979). All of them were rerecording with the same lens format, angle and viewpoint the same locations of Paris, Montreal or Quebec to be able to compare.

** Such visual discussions of the former were for example serials of Sherry Levine on Walker Evans (1981), Christian Boltanski's serials Ten Photographic Portraits of Christian Boltanski,1946–1964, Photo Album of the Family D., 1939–1964, or finally his serial Monuments. Foster called these type of works "subversive" for the very reason of hacking the original author's rights, which subsequently, at the end belonged to the original author, See also: Foster, H. (1985). Subversive Signs. Recordings Art, Spectacle, Cultural Politics. H. Foster. New York, The New Press: 99-121.Fascinating case of Levine's work which ended up as the very work of Walker Evans.

†† The topic was Postmodern in the very sense Jameson formulated in regard to pastiche and parody. See: Jameson, F. (1999). Postmodernism, or, The cultural logic of late capitalism. Durham, Duke University Press.

‡‡ Copy of the negative is used in situations where the negatives have started disappearing. So for example Gertrude Ksebier's, The Road to Rome (I903) has prints made from the copy negative of rephotographed gelatin silver print. See: Bunnell, P. C. (1992). "Pictorial Photography." Record of the Art Museum, Princeton University 51 (2 - The Art of Pictorial Photography 1890-1925): 2+10-15.

§§ But results of recording via dia-positives would be different as they are used when the negatives which are too dense might have been overexposed, overdeveloped or slightly fogged they can be duplicated on the color slide, which would give more contrast.

*** Encyclopedia of XX century photography defines the combination print as "a photograph created through the combining of two or more individual images to form a whole and generally rephotographed to create a seamless final image, as distinguished from montage," contrary to montage which is "An assemblage of photographic images; generally distinguished from collage in both the restriction to photographic images (as opposed to drawn or three-dimensional elements) as well as the presentation of the final image as a unified whole, often by rephotographing. Hannavy, J. (2008). Encyclopedia of nineteenth-century photography. New York ; London, Routledge.

††† In 1990, when this argument was written, digital equipment colours auto adjustment, relying on a very stabile fact of the average "greyness" being 19 percent was still undiscovered. At the time, still the colour definition was largely dependant on RGB rather than b&w definition. Processing of black and white photos on the colour processor was also not fully functional, but was rather giving bluish and greenish results in defining black and gray, while white area was also polluted with the middle value.

Journal of Physics: Conference Series **231** (2010) 012014 doi:10.1088/1742-6596/231/1/012014

International Standards on Stability of Digital Prints

Peter Z Adelstein [1]

Image Permanence Institute at Rochester Institute of Technology, 70 Lomb Memorial Drive, Rochester, NY 14623, USA

E-mail: pzapph@rit.edu

Abstract. The International Standards Organization (ISO) is a worldwide recognized standardizing body which has responsibility for standards on permanence of digital prints. This paper is an update on the progress made to date by ISO in writing test methods in this area. Three technologies are involved, namely ink jet, dye diffusion thermal transfer (dye-sublimation) and electrophotography.

Two types of test methods are possible, namely comparative tests and predictive tests. To date a comparative test on water fastness has been published and final balloting is underway on a comparative test on humidity fastness. Predictive tests are being finalized on thermal stability and pollution susceptibility. The test method on thermal stability is intended to predict the print life during normal aging. One of the testing concerns is that some prints do not show significant image change in practical testing times. The test method on pollution susceptibility only deals with ozone and assumes that the reciprocity law applies. This law assumes that a long time under a low pollutant concentration is equivalent to a short time under the high concentration used in the test procedure.

Longer term studies include a predictive test for light stability and the preparation of a material specification. The latter requires a decision about the proper colour target to be used and what constitutes an unacceptable colour change. Moreover, a specification which gives a predictive life is very dependent upon the conditions the print encounters and will only apply to specific levels of temperature, ozone and light.

1. Introduction

Digital photography has become the dominant technology today for the capture of images. Consequently, colour hardcopies are being printed digitally, and digital imaging includes many more technologies than the chromogenic dyes used in traditional photography. A primary concern in the use of digitally printed materials is their long term stability. The permanence of the image itself is being addressed by the International Organization for Standardization (ISO) while the physical stability is being investigated by organizations such as the Image Permanence Institute. This paper is an update of the current activities of ISO. Although this topic was discussed at the 2003 meeting of the Institute of Physics, there has been considerable progress over the past seven years.

2. International Standards Organization

Before reviewing the ISO work, a brief discussion of this organization is appropriate. ISO is a world-wide recognized standardizing body which publishes standards on testing procedures and specifications. Its scope covers all fields except electrical and electronic standards. Headquartered in

[1] To whom any correspondence should be addressed

Geneva Switzerland, membership is by country, not by organization or by individual. Each country is represented by its respective standardizing body, Great Britain by the British Standards Institution, United States by the American National Standards Institute, Germany by Deutsches Institut fur Normung, etc. The standardizing deliberations are conducted by various technical committees, the work on image permanence being the responsibility of Technical Committee 42. This includes the permanence of photographic film, magnetic tape, optical discs and digital prints. TC42 has a membership of twelve countries and voting is by country, each country having a single vote.

The discussions and deliberations are done in various working groups and task groups. Following general agreement, a proposed standard goes through a series of ballots. This ensures that there is thorough review of the document and all points of view are considered. While this is a time consuming process, it does give confidence that the document represents the best technical information at the time.

3. Standardization of Digital Prints

Standardization of traditional photographic images is relatively uncomplicated compared to digital prints. The former consist of a silver image or a chromogenic dye embedded into a gelatine matrix. While there are differences between manufacturers, these are secondary and the basic technology is similar for all. However, this is not the case with digital prints where at least three different imaging technologies are involved – ink jet, electophotography and dye sublimation.

- Ink jet printers use pigments or dyes which are propelled onto the substrate –including plain, swellable and porous papers.
- Electrophotographic technology uses toners composed of colorants embedded in polymer beads which are transferred to the substrate –typically plain paper– by an electrical charge, then fixed to the paper by heat or pressure.
- Dye-sublimation always uses dyes which are contained in a donor ribbon and transferred to a specialized substrate.

The stability of a digital print not only depends on the applicable technology, but also the type of colorant used to form the image and the substrate the image is printed on.

There is consequently a tremendous number of possible combinations of technologies, colorants and papers. Furthermore, the situation is still more complicated since the life expectancy can be very dependent upon the printer and printer setting that is used. Results obtained using an ISO standard apply only to a very specific digital image, paper and printer.

The primary goal of the ISO standardizing committee is to establish test methods for the different environmental stresses that digital prints may experience. This includes the effects of water, humidity, heat, light, pollutants and abrasion resistance. Two types of test methods are possible, comparative tests between materials and predictive tests. Progress on the test procedures is discussed.

3.1. Comparative Tests

Comparative tests are the simpler and less expensive of the two. They establish testing uniformity between different laboratories and organizations and allow a comparison to be made between materials. However, they give no information about the expected life of the image. Water fastness and humidity fastness are comparative tests

3.1.1. Water Fastness. The only test method that has been completed to date is for water fastness [1]. The digital print is subjected to three different degrees of water exposure. These are a one-minute soak in water without subsequent air drying, a one-minute soak in water followed by blotting, and a one-hour soak in water. The behaviour of the print is rated qualitatively as water resistant, moderately water resistant or not water resistant. It is apparent that this is a highly subjective procedure. Laboratory comparisons indicated that it is not possible to give a more quantitative rating since the degree of water damage is so dependent upon the observer. Nevertheless, this procedure does provide some degree of uniformity in evaluating water fastness.

3.1.2. Humidity Fastness. The resistance of a digital print to humidity can differ from its stability in water. High humidity exposure may cause dyes to bleed or to be transferred to adjacent sheets. Consequently there is a need for a humidity fastness test. Such a test has been under development based on published research [2, 3] and is nearing completion. Tentatively it is proposed to subject prints to (a) 85% RH condition for four weeks, (b) exposure for two weeks at three humidities ranging from 60 to 95% RH depending upon the material and (c) a long time exposure at 20% RH for times up to six months. Evaluation is by colorimetry.

3.2. Predictive Tests

Predictive tests are intended to determine how long a print will last. They are required in order to achieve the goal of preparing a specification for the life expectancy of a digital print. Thermal aging, the effects of pollution and light are all predictive tests. All three involve requirements to evaluate changes in the colour targets resulting from the stress described in the test methods. These colour changes are quite complex. They include the degree of fade of individual colours as well as a neutral colour patch, the change in colour balance caused by unequal colour fading, and the colour increase in the Dmin (white) areas. Measurements are made of density. These predictive tests also require the establishment of end points which determine the degree of image acceptability. These test procedures must be completed within a reasonable time frame. Consequently, this requirement involves subjecting the digital prints to severe stress conditions and then extrapolating the result to a practical condition.

3.2.1. Thermal Aging. The thermal aging test is intended to predict the colour change with time. This prediction is made by the Arrhenius procedure [4] which involves conditioning the print to 50% RH, sealing the print in a humidity proof foil bag followed by incubation for different times at elevated temperatures. Alternatively the specimens can be free hanging in a humidity controlled oven. The incubation time required to reach an end point (such as a specified increase in Dmin or level of fade) is determined. It is then plotted as a function of temperature using the Arrhenius relationship. Figure 1 shows such a treatment which gives a linear relationship with temperature for two hypothetical

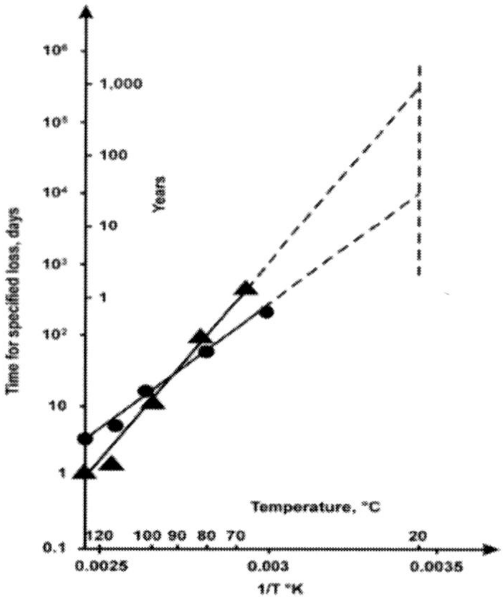

Figure 1. Arrhenius plot in which the end point is reached allowing for predictions.

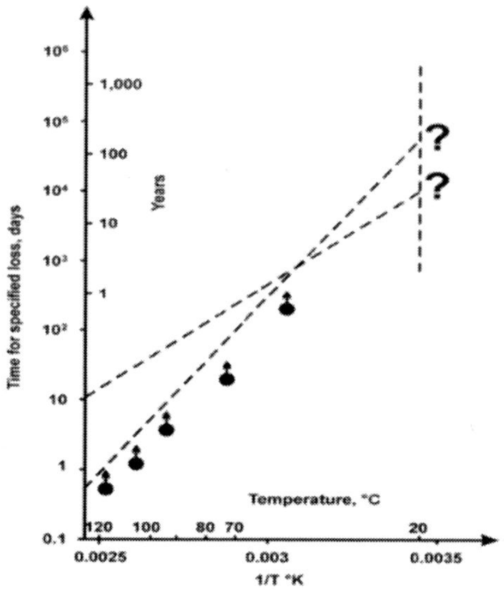

Figure 2. Arrhenius plot in which the end point is not reached not allowing for predictions.

materials and allows a prediction to be made at a practical use temperature. This approach is only applicable if the Arrhenius plot is linear. However, it has been very successfully used to predict the image change of chromogenic dyes and also the life of acetate film base.

An important requirement for the application of the Arrhenius method is that the incubation time/temperature condition must result in the end point being reached. Figure 2 illustrates a test in which the end points were not reached and there is no way in which a prediction can be made. In this illustration comparing the two materials, the one more stable at 120 C is actually less stable at 20 C. The critical criterion is not the stability of the image at the high temperature but the temperature dependence of the end point. One of the test problems found with some digital images is a very high stability, even at elevated temperatures. While this is a distinct advantage, it makes predictions impossible.

A proposed ISO test procedure is now in the final stages of the balloting process.

3.2.2. Pollution Susceptibility. Early testing by the ISO group showed wide variability between laboratories and this was subsequently found to be caused by the effect of pollutants on the image stability, particularly the effect of ozone on dye images applied onto porous paper [5, 6, 7]. The presence of pollutants in the environment is a fact of life. Consequently the standardization of a predictive test to evaluate pollution susceptibility is essential and work has proceeded along this line [8, 9], specifically addressing the effect of ozone which was found to be particularly virulent.

As with the thermal procedure, a test must be completed in a reasonable time. It is essential to subject the digital print to a high stress condition and then extrapolate the results to a practical situation. This is being done by use of the reciprocity law. The assumption is made that a short time at a high ozone concentration is equivalent to a long time at a reasonable ozone concentration [10, 11].

A proposed ISO test procedure is also in the final stages of balloting. This includes standardized methods for producing ozone, measuring the ozone concentration and establishing the relative humidity, temperature and ozone concentration of the accelerated test. In addition, the life expectancy must be based on calculating the results to a standardized practical ozone concentration.

3.2.3. Light Stability. As with the pollution susceptibility test, the light stability procedure also assumes that the reciprocity law applies [12]. The test method requires standardizing the light intensity used to calculate the life expectancy under practical conditions. In addition to the light intensity, the image stability is very dependent upon the wavelength distribution of the exposing illumination. This is obvious by practical experience; sunlight will cause sunburn whereas indoor illumination will not. Currently a proposed procedure is intended to evaluate the image permanence under different lighting conditions. Indoor home lighting conditions are simulated using a xenon arc lamp, while office and public buildings are simulated using fluorescent illumination. Tungsten halogen illumination is also included.

However, there are still a number of unresolved issues. The xenon lamps produces very high heat and temperature control of the test specimen must be established. In addition there are questions about whether a UV filter should be placed in front of the light source. To resolve a number of these issues, a round robin test program is planned in which different laboratories will test identical samples. This will provide information about the reproducibility of the procedure. A proposed final draft is not as close to completion as the thermal and pollution tests and will not be ready in the near future.

3.3. Material Specification

When the predictive tests for thermal stability, pollution susceptibility and light stability have been standardized, the ISO group can then consider the preparation of a specification about the life expectancy of a digital print. The goal would be to specify the number of years that the print would last without exceeding the established end points. However, it must be recognized that it is not possible to ever obtain a single number for life expectancy since the stability of a print differs widely depending whether the pertinent stress is thermal, pollutants or light. While an image printed with a specific technology on one paper may be resistant to one type of stress, such as heat, it may be very susceptible to another, such as pollutants. Moreover, standardizing such specifications involve a number of serious concerns [13, 14].

There are some reservations about the test methods themselves. Each of the stresses evaluated deal with only one stress at a time. It is obviously impossible to prepare test methods which determine all interactions between thermal, pollution and light effects. However, such interactions occur and do represent the real world. Another possible source of error is that both the pollution and the light stability tests assume that the reciprocity law holds. This is a necessary assumption so that the test can be completed in real time, but it is still an assumption.

It is recognized that image stability is very dependent upon the technology involved as well as the nature of the colorants, whether they are dyes, pigments or toners. In addition, the permanence of the image varies with the nature of the paper used. For example dye images on porous paper are very susceptible to ozone attack, whereas on swellable paper they are more resistant. It has also been found that even with the same colorants/paper materials, ink jet stability behaviour can be dependent on the printer used. In summary, any prediction of life expectancy applies only to a very specific material.

There exists another problem which is related to the needs of the user. The end point specification differs for different applications. Obviously the changes that are acceptable for a circus poster are much less stringent than for a fine art photograph

In addition, the end points in the predictive test methods apply to specific levels of stress. The average user does not know what levels of stress apply to a particular situation. While temperature measurements are easily obtained, analysis of the effects of the temperature variation over an extended time period is beyond the reach of most users. The effective temperature requires a calculation since it is not the average temperature which is critical. This is because a low temperature does not undo the harmful effects of an elevated temperature.

While many users may be able to determine the prevailing relative humidity, this may vary over time. Again the harmful effects of a high humidity cannot be compensated by subsequent storage at a low humidity. Consequently the effective humidity must be calculated.

Most users have no information about the ambient pollution levels and no way to determine them. High pollution susceptibility may be extremely important for some collections and of little importance for others. This would be critical in estimating life expectancy in any particular situation.

Another stress level to be considered is the quality of the incident light, specifically the intensity and the wave length. This is unimportant for dark storage but quite important for images displayed at home or in offices.

In summary, the life expectancy values are subject to the uncertainties described above, namely concerns about the test methods, the appropriate end points and the stress levels that are pertinent. Several different approaches have been discussed in the ISO group:

- Not to specify any life expectancy value since the result is open to error. This is technically defensible but does not satisfy the needs of the consumer who desires to have some information about the expected life of the print.
- Provide a life expectancy value in "years" since the consumer requires some knowledge of the property in a unit which is understandable. This has the advantage that the consumer can relate to the life expectancy value. However, it has the disadvantage that the value may be wrong.
- Give a number which represents a rating for life expectancy and which uses a unit other than "years" such as "cumulative dosage" or "nominal years". Such a rating may be useful in comparing prints but it will not give the consumer information about how long the print will still be useable.

This is obviously a complex question which has yet to be resolved.

4. Conclusion

Preparing standards on the life expectancy of digital prints is extremely complicated. Some of the important variables were not realized when this activity was started and this obviously delayed progress. However, there have been substantial contributions in recent years. To date one comparative standard on water susceptibility has been completed, and the final balloting progress is underway for another comparative test method on humidity susceptibility. Two predictive test methods on thermal and pollution susceptibility are also in the final stages of balloting. Unresolved issues remain on a light stability test method and on a material specification.

References

[1] *ISO 18935 Imaging materials – Colour images on paper prints – Determination of indoor water resistance of printed colour images*

[2] Kaimoto H, Shibahara Y and Shizuoka F 2007 Test Methods of Humidity Fastness of Inkjet Printing Materials *Technical Program and Proc. of IS&T NIP23* (Anchorage, Alaska: The Society for Imaging Science and Technology) pp 728-731

[3] Hofmann R and Reber J 2009 Humidity Diffusion of Dye-based IJ Prints *Technical Program and Proc. of IS&T NIP25* (Louisville, Kentucky: The Society for Imaging Science and Technology) pp 126-129

[4] *ISO 18924 Imaging materials – Test method for Arrhenius-type predictions*

[5] Oakland M D, Bugner R, Levesque R and Vanhanehem R 2001 Ozone Concentration Effects on the Dark Face of Inkjet Photographic Prints *Technical Program and Proc. of IS&T NIP17* (Ft. Lauderdale, Florida: The Society for Imaging Science and Technology) pp 175-178

[6] Wight P 2002 Features of Reciprocity Failure and Ozone Fading in Ink Jet Images *Technical Program and Proc. of IS&T NIP18* (San Diego, California: The Society for Imaging Science and Technology) pp 334-336

[7] Moeller S, Kaumkoetter D, Schmidt W and Papier G A Review of the Evolution of InkJet print Durability Against Environmental Gases *Technical Program and Proc. of IS&T NIP23* (Anchorage, Alaska: The Society for Imaging Science and Technology) pp 755-758

[8] Reber J and Hofmann R 2006 Correlation of Ozone Test Chamber Data with Real Life Permanence of Inkjet Prints *Technical Program and Proc. of IS&T NIP22* (Denver, Colorado: The Society for Imaging Science and Technology) pp 231-234

[9] Comstock M and McCarthy A 2008 Effect of Temperature and Humidity on Rate of Image Fading in Ozone *Technical Program and Proc. of IS&T NIP24* (Pittsburgh, Pennsylvania: The Society for Imaging Science and Technology) pp 237-241

[10] Bugner D, Van Hanehem R, Artz P and Zaccour D 2003 Update on Reciprocity Effects for Accelerated Ozone Fade Testing of Inkjet Photographic Prints *Technical Program and Proc. of IS&T NIP19* (New Orleans, Louisiana: The Society for Imaging Science and Technology) pp 397-401

[11] Comstock M and McCarthy A 2009 Further Studies on Reciprocity Effects for Accelerated Ozone Tests Compared to Ambient Air Exposure *Technical Program and Proc. of IS&T NIP25* (Louisville. Kentucky: The Society for Imaging Science and Technology) pp 134-137

[12] Bugner D and Lindstrom B 2006 Further Investigations into Accelerated Light Fade Reciprocity of Inkjet Photographic Prints *Technical Program and Proc. of IS&T NIP22* (Denver, Colorado: The Society for Imaging Science and Technology) pp 235-238

[13] Mason P 2008 Accuracy in Photo Print Life Prediction *Technical program and Proc. of IS&T NIP24* (Pittsburgh, Pennsylvania: The Society for Imaging Science and Technology) pp 242-246

[14] Image Permanence Institute 2009 *A Consumer Guide to Understanding Permanence Testing*

JPEG 2000 standards in digital preservation

Richard Clark

Elysium Ltd and JPEG Honorary Webmaster
Milton House, Whitehill Road, Crowborough, East Sussex TN6 1LB

richard-iop@elysium.ltd.uk

Abstract. Although JPEG 2000 is now a preferred format for archival image preservation, its use still requires care and consideration. This paper addresses how some of these issues can be addressed from a pragmatic perspective, and also looks at how the technology and file formats are developing to provide a full and consistent architecture for handling and searching large visual archives. New methods of addressing perceptual quality are discussed, showing that JPEG 2000 offers the best quality compression, although the metrics indicate that lower compression rates are recommended for preservation formats than might be assumed from previous simple metrics such as PSNR.

1. Introduction

Many large archives are now considering the move to all digital operation, with substantial collections already digitised. Many major collections are following the lead of the US Library of Congress [1] and the National Archives of Japan [2] in adopting JPEG 2000 as the preferred format for long term storage, including:

- the Wellcome Trust's collection of early documents on the history of medicine, and other scientific collections [3]
- the British Library's newspaper collections [4]
- The National Archives' massive document collection [5]

The rationale behind the choice of JPEG 2000 has been varied, but the volume of images in each case is massive, and a prime motivator is often the saving of disk space, of the order of petabytes. There have been many studies of the usage of JPEG 2000 in archiving, including the author's own presentation to the Digital Preservation Coalition [6], and the authoring of a "Guide to the practical implementation of JPEG 2000" [7]. In more recent years, studies for the University of Connecticut [8], the Library of Congress [1], Library and Archives Canada [9], and most recently the Wellcome Trust (through consultancy from Buckley and Tanner [10]) have added to the wealth of information available on the internet and in books.

Although many tools now exist, both commercially and in the Open Source arena, their quality and adherence to the various parts of the standard vary significantly. The unwary can find that converting volumes of images in bulk as part of a production workflow is very different from carrying out a few simple tests to show that JPEG 2000 can deliver compression gains.

Nevertheless, the JPEG committee's own Historical Archive (which has collected some 7000 or so documents dealing with the technology used by its members) shows in a microcosm some of the problems that can occur. Most documents have had to be converted to Word 97 or later, with no guarantee that later versions will render correctly items such as embedded images. Although Word 5.5 is still available as a freely downloadable DOS package [11], it will not run under 32 bit operating system like Vista.

Early PDFs in the collection were created with version 1 of Acrobat - these seem to be less problematical than Word documents for now, and do render well in the latest version of Acrobat. The original JPEG standard itself was edited in WPS plus on a DEC VAX computer, no longer an option. Many original paper documents are in manuscript, or annotated, and any format for preserving the archive needs to be capable not only of rendering the original as truthfully as possible, but also should be able to handle the associated metadata in an unambiguous and searchable format. For text, and particularly printed text, this may not be much of a problem - however the digitisation process and metadata capture even for a tiny archive such as this has not been problem free.

At a time when operating systems are upgrading almost every year, hardware platforms change dramatically, and many formats are proprietary and often undocumented, keeping tools to decode and render images digitally has to rely on a stable, documented and standardised approach if it is to stand any chance of success. It must also provide some mechanism to take into account the plethora of copyright and IPR issues that can encumber the collection and distribution of contents. The recent history of the Domesday project on the BBC microcomputer (and the lack of relevant standards at the time it was created) at least shows this [12].

In this paper we will look at how tools are being created to allow users to work better with image data, and how the work carried out for JPEG 2000 has led to many spin-off developments, improving the possibility for automation in the future.

2. JPEG 2000 as an architecture

The standardisation process for JPEG 2000 has taken place over nearly 14 years - with much of the later effort concentrated on using the core coding technology to deliver novel or practical implementations of the underlying architecture. The standard is currently in some 13 parts [13], covering such application domains as :

- basic archival needs
- handling wide varieties of embedded metadata
- for moving images, linked in with the same file format [14] that is used by MPEG in its MPEG 4 standard - this is now the accepted method for coding movies for distribution to cinemas, using standards based on JPEG 2000 from the Digital Cinema Initiative
- increasingly, as a part of other component architectures such as the medical sector's work on DICOM [15], or in geocoding [16], [17]
- for handling very large images - for instance those generated in space exploration or satellite imaging, or in some cases for digital preservation (for instance of large maps within the Japanese National Archives [2]) - although at the distribution and viewing level, JPEG 2000's JPIP protocol faces strong competition from alternatives such as IIPImage [18] or Photosynth [19]. JPEG 2000 can still be used as a backend storage format for these distribution methods, as JPEG 2000's ability to be effectively transcoded offers many benefits. The Old Maps Online site, developed by the Moravian Library is an excellent example of what can be achieved by mixing technologies using JPEG 2000 as a base [20].
- new developments in securing images are possible using JPSec - a JPEG 2000 derived security architecture allowing images to be protected with different features for example at different resolution levels
- applications using wireless or 3D imaging are starting to emerge

Nevertheless, in two key areas, it seems likely that JPEG 2000 will struggle for market share. In the Internet uncertainty over the legal position of the many patents covering aspects of JPEG 2000 technology (and a track record of contentious litigation already over GIF and subsequently JPEG) has made many players uneasy. The JPEG committee made every effort to ensure that a final baseline solution could be offered with all participants offering free licensing - however without payment, it has proven impossible to actually source appropriate licenses from the 20+ organisations who made these generous offers. Although to date, the author is unaware of any litigation over JPEG 2000 patents (unlike the situation with JPEG), nevertheless, the value of the market (and the enormous revenues generated from patent pools in the video and audio MPEG arena) tend to indicate that there is a particular risk to the largest organisations (such as Microsoft, Apple, Adobe and so on) in investing in any new multimedia technology with global aspirations.

The other area in which there has been very little sign of activity has been the digital camera market. Although originally a key potential market, the proliferation of camera RAW files and an interest in scene referred rather than output referred imagery has meant the target market is much more limited than originally envisaged. The cost of memory has fallen to the extent that most professional photographers will shoot in RAW plus JPEG simultaneously, whilst battery life is still an important issue, and one where the use of JPEG 2000 carries a significant penalty against original JPEG, and which impacts the all important consumer end of the market. The introduction of Microsoft's HD Photo as JPEG XR is probably another killer for this market - offering (one hopes) an improved patent landscape, implementation in many of the core products from a market leader (and prospects of its use in browsers therefore). It is hard, therefore, to develop a good rationale for why JPEG 2000 should be preferred as a next generation compression standard for cameras. Test results indicate that JPEG 2000 can compress better visually than JPEG XR [21], and its coding architecture seems to have many benefits. Nevertheless, practical issues rule, and it seems therefore that where JPEG 2000 is best is in underpinning other imaging developments, and therefore as a basis for archival preservation of images. The remainder of this paper studies how its use is best addressed.

3. Key issues

JPEG 2000 codestreams can be created by a variety of different software packages. In the Open Source arena there are options such as Jasper [21] and OpenJPEG [22] (which took the work of both Jasper and JJ2000 as part of JPEG 2000's reference software as a base, and has extended both significantly). There are many vendors including JPEG 2000 capabilities in their toolkits, including Luratech, Aware, KakaduSoft, Accusoft, Leadtools, Adobe, Apple, and others. In some cases (for example Apple's apparent use of Kakadu as a base for its codecs) results are very similar. In others there are significant differences in how well the toolkit works in such areas as :

- memory usage. JPEG 2000 images can be very large, and how memory is handled can make big difference to performance
- colour transcoding. JPEG 2000 allows for the use of both enumerated and embedded ICC profile information - however the range is different to that documented by ICC for example
- performance - large workflows can require very fast automated processing of images - the National Archive for example are looking at using hardware encoding based on technology developed for the digital cinema industry
- metadata handling - there are many complementary (and sometimes conflicting) metadata standards. The publishers seem to be coming down in favour of a derivative of Adobe's XMP, whilst the archiving community use Z39.87. The JPEG 2000 JPX file formats suggests DIG35, and the use of EXIF in digital cameras is very widespread - however mostly using proprietary 'Makernotes' to extend the specification, and occasionally adding encryption to ensure security (and perhaps sales of proprietary processing software!)
- platform - most software toolkits target Windows, although Apple Mac computers and Photoshop or Lightroom usually form an initial part of the production chain
- import capabilities - most will accept a wide variety of input formats, and in particular TIFF. However it is often far from clear as to how aspects of the complex TIFF file structure will be handled, such as issues of colour space or metadata import, and there are no rela standards or implementor's guides that offer 'best practice' advice in this area

This section will now look at some key issues in more detail.

Journal of Physics: Conference Series **231** (2010) 012015 doi:10.1088/1742-6596/231/1/012015

3.1. Measuring image quality

Despite recognition that PSNR (effectively a measure of the distortion introduced against an original image) is not very good as a measure of image quality, it continues to be the dominant cited metric when coding schemes are compared. Very often, in the high quality preservation arena, images are pre-processed in products such as Adobe Photoshop to give a 'good' image quality. This can introduce very significant distortion - proponents of SSIM [24] as a better measure often cite examples such as that shown below. Here the well known 'Lena' image [25] is taken, and improved by correcting the colour balance and sharpening the image slightly. Given the original was scanned from the November 1972 Playboy magazine around 1973, using fairly basic technology on a screened and magazine printed photo, it is hardly surprising that a little image processing improves the perceptual quality fairly dramatically. This however gives a fairly low PSNR reading against the 'original', showing high levels of distortion, as might be expected. Using PSNR as a metric, the blurred image shown below is almost 6dB 'better' in quality than the 'improved' one - whilst the heavily pixellated image is almost exactly the same quality using the metric.

Within the JPEG committee, Dr Thomas Richter has been very active in developing the work of Wang and others on a 'Structural SIMilarity index' (SSIM) into a variant he has called Multiscale SSIM (MSSIM), which appears to give a much better correlation between perceptual quality of an original and a modified image. As indicated in Figure 1, the MSSIM scores show a much closer relationship to the visually perceived quality than does the PSNR.

As a result, a comparison can be drawn between the quality of JPEG 2000, and of JPEG at varying compression ratios. The results presented use a high quality 8 bit per pixel image (Figure 2) of a picture by Ingres scanned from an original in the Louvre [26], as typical of the sort of content that might require care and attention in its preservation. As can be seen, the picture is already deteriorating and hence it is very valuable to conservationists to have a range of quality images taken at various stages during a restoration project - often using other forms of illumination or sensors, such as raking light, infra red, and x-ray photography. The image is 4088 pixels wide by 4608 deep.

The image has been compressed using the Independent JPEG Group JPEG codec [27], and using the Kakadu JPEG 2000 codec [28] at various rates. These are then measured against the PSNR metric and against MSSIM (using the tool as supplied to the JPEG committee by Thomas Richter [29]). In addition, the values have been provided for PSNR and MSSIM for a sub-sampled image, which has been sharpened with sampling using Lanczos resampling to try and give a good quality interpolation. As is clearly seen in the results, JPEG and JPEG 2000 compression is far more effective as a control of the final image size for archiving than lowering the sampling rate, measured by both PSNR and MSSIM measures of quality - in fact simply halving the sampling rate (giving a quarter size for storage purposes) is roughly equivalent to compression rates of around 50:1. This is a strong pointer to the need to keep scanners working at an optimum and high resolution, and applying lossy compression, rather than simply believing that lossless compression at a given resolution is in some way better than lossy compression at a higher resolution. Essentially, the so-called 'lossless' compression is throwing away a lot more data than would be the case if a higher scan resolution was used with lossy compression.

Journal of Physics: Conference Series **231** (2010) 012015 doi:10.1088/1742-6596/231/1/012015

 Original	 'Improved' version PSNR - 22.6 dB MSSIM - 40.4
 Blurred version PSNR -28.2 dB MSSIM - 21.0	 Pixellated version PSNR - 22.3 dB MSSIM - 10.6

Figure 1 - Comparison of PSNR and MSSIM as metrics

Figure 3 shows the curves for PSNR as a measure, which tend to show at a first glance that compression benefits flatten off at around 2 bits per pixel - a compression ratio of around 12:1, and also that JPEG 2000 is significantly better than JPEG compression in its performance. However, using MSSIM as in Figure 4, the picture shifts significantly - now optimum compression is around 3 bits per pixel - a compression ratio of only 8:1, and with relatively little performance difference between JPEG and JPEG 2000. I believe this is a fairer reflection of the real situation - interestingly both Buckley [10] and Desrochers [9] come up with similar recommendations for the practical use of JPEG 2000, with a target compression of around 8:1.

Figure 2 - Image used for quality comparison [26]

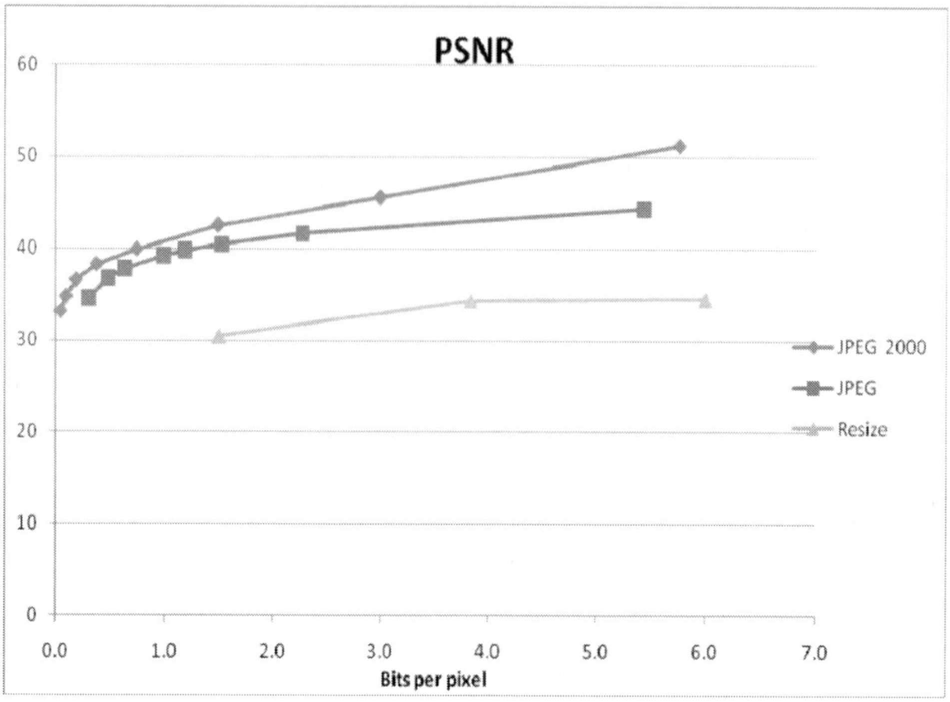

Figure 3 - Measured PSNR for compressed versions of Figure 2

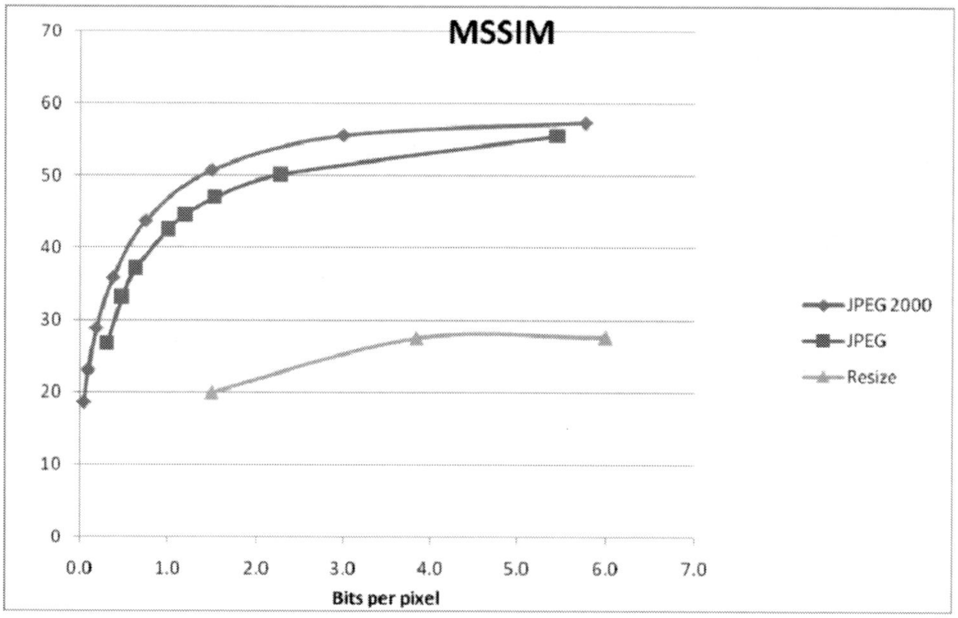

Figure 4 - Measured MSSIM of compressed versions of Figure 2

Needless to say the picture is not as simple as painted in this brief paper, as there are techniques using visual weighting which can be applied to JPEG 2000 to improve its visual performance (and measured performance using MSSIM) - however there is also a counter argument which indicates that applying improved entropy compression or variable quantisation (based on an SSIM metric) techniques to JPEG can also improve results, and so the results are probably a fair reflection. JPEG 2000 wins not because it is better at compressing images (although undoubtedly it is), but because of the amount of thought and specification that went into its development, and its potential use as a single codestream that can easily deliver a wide range of qualities and resolutions by transcoding without the need to preserve multiple master copies.

It is important in any JPEG workflow system for digital preservation to look at whether some kind of quality system can be applied automatically to ensure that compression has not introduced undue distortion. It is suggested that MSSIM forms a much more effective tool for this purpose than a traditional and simpler measure such as PSNR.

3.2. File format

JPEG 2000 defines two different file formats - JP2 and JPX. JP2 is a basic file format, defined and required even in baseline implementations. Typically files in this format are suffixed **.JP2.** JPX extends this significantly, and is defined in the second part of the JPEG 2000 standard. Although the formal definition for JPX and its MIME type definition in RFC 3745 [30] recommends that files should be suffixed **.JPF**, in practice many implementors have ignored this and use the suffix **.JPX**.

The definition of both file formats is beyond the scope of a brief paper, and in any event is well covered in the standard and elsewhere. Both formats allow properly defined XML data to be freely embedded within a file, however JPX also suggests a metadata syntax (based on the work in I3A on the DIG35 standard [31]), and permits multiple image codestreams within one file.

A major difference between the two from a preservation perspective is the way that they handle colour information. Formally, JP2 files can either contain a simple enumerated ICC profile (restricted to sRGB [32], grey scale, and sYCC [33]), or a 'restricted' ICC profile based on either a three channel matrix based or monochrome profile. In practical terms this means that Adobe RGB 1998 [34] and Pro Photo [35] can only be indicated by their profile, not by a simple enumeration, and CMYK based profiles cannot be handled as JP2 files. JPX files can contain full ICC profiles, and also have a much wider range of enumerated values. The decision was however taken to restrict these to internationally standardised profiles, and as a result there is poor practical correlation between the pragmatic range of profiles maintained by ICC at http://www.color.org, and those enumerated in the standard. Profile embedding seems to be the way to go from an accuracy perspective - the standard even allows multiple profiles to be embedded allowing for implementation dependent handling. The suspicion is that this has confused the market somewhat - certainly designers need to be very aware of how specific implementations handle colour mapping, and in particular the conversion from an input file format such as and Adobe Photoshop TIFF [36] file with an embedded profile (often Adobe RGB 1988 [34]). For 8 bit per component files, there is of course a good argument to suggest that sRGB is an effective solution, marrying bit depth to the lower range of colours it handles to reduce quantisation effects. In the real world however, with scanners and cameras offering RAW files [37] with accurate bit depths of 14 bits or beyond, another approach is necessary, and here it appears that ProPhoto (as ROMM RGB [35]) may be a better way to go, particularly as it matches native RAW processing within Photoshop, and is also available as an enumerated file format within JPX.

3.3. Metadata options

As noted in 3.2, both file formats supported for JPEG 2000 can handle XML metadata, simply by including the information in a XML 'box'. In addition, both file formats permit the inclusion of other forms of binary data in different box type, defined with a unique identifier (UUID). In practice significant amounts of metadata are typically included in this way - for example the EXIF metadata captured by a camera. JPX also includes fields such as a 'licence plate' which can hold ownership or IPR information - often replicated in metadata held in XML or UUID boxes. Given that many variants of metadata may already exist, and could potentially be encapsulated in a JP2 or JPX file, care must be taken when editing or updating the information, as the problem already apparent with many JPG files on the Internet still exists - a user may think they have the correct metadata when using specific display tools, only to find out that there was additional information held elsewhere in the file. A good example of this is some of the copyright information in images made available in some versions of Microsoft Windows as sample images - here information held in Photoshop specific application segments indicates a copyright holder - however other metadata (and perhaps the folder title) might lead users to assume that no such copyright exists or is being claimed. As visual search tools improve (and the power of litigation), this has proven for many small users to be an expensive oversight.

In a further twist, many input routines take TIFF files from products such as Adobe Photoshop or Lightroom or Apple's Aperture, which can include undefined Makernote information from digital camera manufacturers. As TIFF files can contain absolute pointers to items in the file, it is possible that transcoding these into an XML box in JP2 or JPX may break these links. Until some rationale emerges in this marketplace, it seems that this type of problem is set to continue - however it is certainly an issue that needs to be addressed in the context of preserving as much information as feasible from 'born digital' imagery.

3.4. Compressed or Lossless?

The argument for using lossless image compression (which JPEG 2000 can support in full) does seem to be weakening - evidence such as that shown in Figure 3 & 4 above indicates that there is little if any technical benefit to using lossless compression, particularly if using lossy compression can mean starting with a higher resolution original. Practical tests, such as those carried out by Buckley [10] also confirm these findings, and most practical workflows are now optimised to provide more speed and flexibility by taking advantage of lossy compression optimisations in JPEG 2000.

3.5. Codestream structure

One of the key, if not the key benefit of using JPEG 2000 is the way that the codestream of an image can be ordered into small packets, which can then deliver a wide range of features including regions of interest, multiple quality levels, multiple resolutions, and some optimisations which can help encoding or decoding, such as the use of tiling. These are affected by the order in which thee packets are assembled, which can prioritise quality, resolution, or even individual components. Buckley [10] recommends RCPL ordering, and this has obvious benefits for web style applications, where images need to be generated at several different resolution levels rapidly. Again, the benefits of the different ordering schemes are well covered both in the standard, and in software tools and books [38], [39].

3.6. Tiling

Using tiles can improve the performance of the codec, as for very large images, memory storage becomes an issue. For some examples (for example streaming images directly from a satellite) tiling is the only option. The impact of tiling is minimal visually or in compression terms, and it can improve access to sections of an image. Some researchers recommend the use of tiles, typically at either 512 by 512 or 1024 by 1024, although in most cases this appears to have been because the actual codec in use performed better with tiles rather than out of any real benefit. Although tiles may introduce some edge effects at high compression ratios, these will be unnoticeable at the sort of compression ratios used for digital preservation, and hence tiling (say at 1024 by 1024) remains a valid option to consider for the final format.

3.7. Validation

Recognising that many encoders are not perfect, and that options exist when transcoding that may not be visible to the user, JHOVE [40] has been created as a validation and identification tool for the Harvard University JSTOR initiative [41]. This can be used both to identify formats, and to ensure that they comply with specific profiles. JPEG 2000 is included, and its use in archival work flows is a valuable addition in ensuring that formats have been correctly generated. Figure 5 (from the JHOVE web site http://hul.harvard.edu/jhove/index.html) shows how JHOVE can be fitted into a workflow for digital preservation.

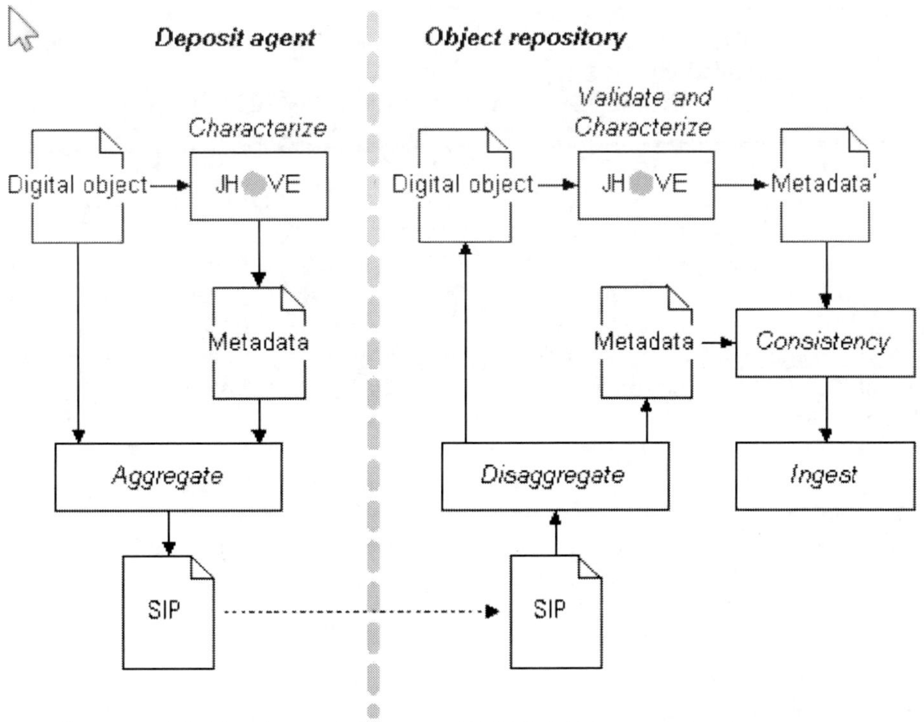

Figure 5 - Use of JHOVE in validation of JPEG 2000 converted images

A possible further option would be to include a feedback loop measuring the accuracy of compression immediately prior to JHOVE validation in the object repository. As suggested by the work for this paper, using MSSIM in this respect seems to offer promising results.

4. Conclusions

This paper has provided a brief introduction to the value of, and problems associated with adopting JPEG 2000 as a preferred format for digital image compression in preservation and archiving. It shows that using tools such as MSSIM, which attempt to apply perceptual quality values to measuring image quality are both more accurate, and more useful in the context of workflow methods for image asset management.

Although JPEG 2000 does offer a valid and effective tool for storing long term image archives, its use does need to be carefully considered, and tools that are applied to the conversion process need to be properly evaluated and configured before committing to large scale production. Metadata and colour space handling in particular require particular and ongoing consideration, and although JPEG 2000 file formats can handle a wide variety of options, the actual file format in use needs to be tightly specified, and dynamically linked into databases used for digital asset management.

References

[1] http://memory.loc.gov/ammem/help/mrsid.html
[2] http://www.digital.archives.go.jp/gallery/view/category/categoryArchivesEn/0200000000/default
[3] http://www.bl.uk/aboutus/stratpolprog/ccare/introduction/digital/digpresmicro.pdf
[4] http://wellcomelibrary.blogspot.com/2009/09/wellcome-library-to-use-jpeg2000-image.html

[5] Private discussion, The National Archives of the UK
[6] Clark, Richard (2007) "Update on JPEG 2000 and JPEG 2000 activities" DPC/BL Joint JPEG 2000 Workshop, 25th June 2007 http://www.dpconline.org/technology-watch-reports/download-document/158-update-on-jpeg-2000-and-jpeg-activities-clark.html
[7] Clark, Richard and Colyer, Greg, (2003) *"PD 6777:Guide to the Practical Implementation of JPEG 2007"*. (Great Britain, BSI)
[8] Lowe, David and Bennett, Michael J., "A Status Report on JPEG 2000 Implementation for Still Images: The UConn Survey" (2009). UConn Libraries Published Works. Paper 19. http://digitalcommons.uconn.edu/libr_pubs/19
[9] Desrochers, P. and B. Thurgood, JPEG 2000 Implementation at Library and Archhives Canada, in J. Trant and D. Bearmans (eds.) *Museums and the Web 2007: Proceedings,* Toronto: Archives & Museum Informatices, published March 1, 2007 Consulted March 24, 2010
[10] Buckley, Robert and Tanner, Simon, "JPEG 2000 as a Preservation and Access ormat for the Wellcome Trust Digital Library" (2009). library.wellcome.ac.uk/assets/wtx056572.pdf
[11] http://download.microsoft.com/download/word97win/Wd55_be/97/WIN98/EN-US/Wd55_ben.exe
[12] http://en.wikipedia.org/wiki/BBC_Domesday_Project
[13] http://www.iso.org/iso/search.htm?qt=15444&searchSubmit=Search&sort=rel&type=simple&published=on
[14] http://www.iso.org/iso/iso_catalogue/catalogue_tc/catalogue_detail.htm?csnumber=51537
[15] ftp://medical.nema.org/medical/dicom/final/sup61_ft.pdf
[16] http://en.giswiki.net/wiki/GeoJPEG2000
[17] http://www.opengeospatial.org/standards/gmljp2
[18] http://iipimage.sourceforge.net/
[19] http://photosynth.net/
[20] http://help.oldmapsonline.org/jpeg2000
[21] http://www.ece.uvic.ca/~mdadams/jasper/
[22] http://www.openjpeg.org/
[23] http://www.iso.org/iso/iso_catalogue/catalogue_tc/catalogue_detail.htm?csnumber=51609
[24] http://www.ece.uwaterloo.ca/~z70wang/research/ssim/
[25] http://en.wikipedia.org/wiki/Lenna
[26] Scan of Portrait of the Italian Composer Cherubini and the Muse of Lyrical Poetry By: Jean Auguste Dominique Ingres (1780 - 1867) ... Location: Louvre, Paris
Provided to JPEG committee for evaluation purposes in 1995 as original TIFF image.
[27] http://www.ijg.org/
[28] http://www.kakadusoftware.com/
[29] Richter, Thomas TU-Berlin, "*Multiscale SSIM software*". Contribution HDPn0006, May 2007, JPEG committee document register (http://www.jpeg.org)
[30] http://tools.ietf.org/html/rfc3745
[31] http://members.digitalimaging.org/shareddocs/downloads/dig35v1.0-Sept00.pdf
[32] http://www.color.org/chardata/rgb/srgb.pdf
[33] http://www.color.org/sYCC.pdf
[34] http://www.color.org/chardata/rgb/adobergb.xalter
[35] http://www.color.org/ROMMRGB.pdf
[36] http://partners.adobe.com/public/developer/en/tiff/TIFF6.pdf
[37] http://en.wikipedia.org/wiki/Raw_image_format
[38] Schelkens, Skodras, Ebrahimi 2009 *The JPEG 2000 Suite* (Great Britain, Wiley)
[39] Taubman, David S. and Marcellin, Michael W. (2002) *JPEG2000: image compression fundamentals, standards, and practice* (USA, Kluwer Academic Publishers)

[40] http://hul.harvard.edu/jhove/index.html
[41] http://www.jstor.org/

Acknowledgements
The work in this paper is based on discussions and advice from many colleagues and friends within the JPEG community and elsewhere. The author would specifically like to thank Thomas Richter, Rob Buckley and Phil Green for their valuable help and assistance over many years, and Thomas in particular for creating the code used to measure MSSIM as presented.

Standards for illumination of digital prints and photographs

Phil Green

London College of Communication, Elephant and Castle, London SE1 6SB

pj.green@lcc.arts.ac.uk

Abstract. Standards for illuminating digital prints and photographs have a number of quite different applications. In the graphic arts industry, the main applications are defined as appraisal and critical comparison, for which 500lux and 2000lux are specified in ISO 3664. In the museum world much lower levels of illumination are imposed when artefacts are considered to be prone to damage from such exposure. For display and storage of photographic prints, BS 5454:2000 is applicable and specifies maximum levels of 50 lux and 200 lux respectively. While these standards provide recommendations for exposure to radiant energy with the goal of limiting damage to materials and maximising visual discrimination, there is a need for more data on the radiative damage spectrum for the materials used in digital prints and photographs and other artefacts, and on the viewing conditions which can maximise visual performance for specific tasks. It is recommended that radiative exposure is measured in watts per square metre instead of lux to give a better indication of the propensity for radiative damage of a given illumination source.

1. Introduction

Viewing conditions for the display of reflective artefacts such as digital prints and photographs have three independent and potentially conflicting goals: first, to display the prints which is maximally satisfying for the viewer and which enables the artistic intent of the creator to be conveyed; secondly, to minimise the radiative damage which might arise in the viewing environment; and thirdly to achieve the most sustainable display possible by minimising the energy required in the display.

2. Performance on visual discrimination tasks

User satisfaction is difficult to quantify directly, and therefore it is preferable to measure performance on visual discrimination tasks under different lighting conditions. The ability to perform a visual discrimination task is of course dependent on characteristics of the human visual system, such as spectral sensitivity, luminance-level adaptation, and contrast sensitivity. It is also possible to evaluate lighting conditions by measuring observer preferences for reproductions under the different conditions, although this approach introduces additional elements of subjectivity.

From the literature on psychophysics (including the Weber-Fechner Law and Steven's Law [1]) it can be deduced that the perceived difference between two reflective stimuli increases as the incident illumination increases. However, relatively little work has been done to test this assumption systematically under real viewing conditions. It has recently been reported [2] that observer performance in detecting a 3 CIELAB ΔE^*_{ab} threshold difference actually fell as illuminance increased, and the optimum illuminance for this threshold detection task was approximately 600 lux. It is not clear why performance fell at high illuminance levels in this case, but it is possible that glare on the prints influenced visual perception despite the notional geometry of the viewing booth being designed to minimise this possibility.

The quality of a lighting source for visual discrimination is usually measured by the CIE Colour Rendering Index [3] (CRI), which for a set of standard reflectances compares the visual response under a reference and test illumination spectral power distribution. A CRI of 100 indicates a perfect match, while a CRI > 90 indicates a good match. The CRI does not indicate the absolute ability of a human observer to discriminate between two stimuli, but the predicted closeness of the match under two different viewing conditions; so for example a tungsten source may have a CRI close to 100 relative to the reference CIE Illuminant A, but observers may perceive threshold colour differences better under an illumination source with a higher correlated colour temperature.

The visual discrimination performance required in different display applications varies considerably, from basic recognition of high-contrast shapes, through perception of tonal differences in monochrome photographs and documents, recognition of basic colour categories, colour preference judgements, through to fine discrimination of small tone and colour differences. This suggests a finer granularity in the specification of viewing illuminances, and further research to support such specifications. For certain tasks (such as viewing historic documents) it may not even be necessary to have illumination at the photopic level, which suggests adapting luminances below 10 candelas per square metre (cd m^{-2}) may be feasible.

3. Light sources, efficacy and visual performance

Current drivers of new developments in lighting technology today include the environmental imperative for greater efficacy, and the potential for improved colour rendering quality. For the purpose of viewing photographs and prints we can now add commercially-available LED light sources to the traditional incandescent and fluorescent lamps.

Although fluorescent lamps have a spectral emission characterised by narrow-band spikes, by adjusting the phosphors used to coat the tube manufacturers are able to produce fluorescent lamps with a high CRI relative to CIE daylight illuminants. Currently lamps with a CRI of 98 and above relative to CIE D50 are commercially available [4]. Fluorescent lamps have efficacies of up to 50 lumens per watt (lW^{-1}), compared with around 15 lW^{-1} for incandescent.

LEDs tend to have more broadband emissions, and white LED sources are manufactured from a combination of two or more LEDs to give complete coverage of the visual spectrum. In some systems the output of three or more LEDs can be independently controlled to modulate the overall spectral power distribution, although it is still difficult to match to reference illuminants and CRIs are below 90.. Although LEDs have a greater apparent efficacy (achieving over 50 lumens per watt in commercially-available packages and reportedly over 200 lW-1 in the lab [5]), they produce a more directional emission which requires diffusion for illumination of photographic displays, and a consequent loss of efficiency. Commercially available LEDs also suffer from a loss of efficacy as current increases.

As yet LED lamps have made little impact in the photograph display and conservation fields. For both fluorescent and LED sources, the spectral power distribution differs very considerably from traditional incandescent sources, and the lux measure used to quantify illuminance does not adequately convey the propensity for radiative damage in such cases, as is shown below.

Fluorescent lamps have several sharp peaks in their relative spectral power distribution. Most such lamps exhibit strong peaks at approximately 405 and 437nm, arising from the emission of the mercury vapour of the lamp. 'White' LEDs used as illumination sources are commonly InGaN blue LEDs with a strong peak at approximately 465nm, combined with a phosphor coating to generate emission in the green and red regions.

It is possible that the presence of these strong peaks in the short-wavelength region of the visual spectrum in LED and fluorescent lamps may accelerate radiative damage of certain materials.

4. Standards for viewing photographic prints

For viewing photographs and prints, applicable ISO standards that are published or in development are:

- ISO 3664:2009 Graphic technology and photography — Viewing conditions [6]
- BS 5454:2000 Recommendations for the storage and exhibition of archival documents [7]
- ISO 18937:2010 (WD) Imaging materials — Photographic reflection prints — Methods for measuring indoor light stability [8]

4.1. ISO 3664:2009

ISO TC 42, working jointly with ISO TC 130, has recently revised the graphic arts viewing standard ISO 3664, which defines viewing conditions for practical appraisal and critical comparison of prints and other media. This new revision provides a clearer definition of the reference illuminant and the procedures to be used to evaluate simulators of that illuminant. Such a definition is of great importance in graphic arts, since the determination as to whether a reproduction is a good visual match to a proof or original requires a precisely standardised viewing condition to avoid the possibility of false matches or false rejections.

The reference illuminant is defined to be the CIE D50 daylight illuminant, whose spectral power distribution is given over the range 300-780nm. Illumination levels for critical comparison (P1) and practical appraisal (P2) conditions are defined as 2000 lux and 500 lux respectively, with tolerances of 500 and 125 lux.

As ISO 3664:2009 is concerned with optimising the appearance of hard copy prints and photographs, it specifies that the viewing illumination should have the same effective UV content as that of the D50 reference illuminant, in order to achieve the higher brightness and whiteness levels intended by the manufacturers of the substrates used.

In addition to illuminance and uniformity specifications, ISO 3664:2009 defines the performance of viewing illumination in terms of three indicators: the CIE Colour Rendering Index, the CIE Metamerism Index for the visible spectrum (MI_{VIS}), and the CIE UV Metamerism Index (MI_{UV}). The MI values are based on the predicted colour difference between two standard reflectances which match under certain SPDs, when illuminated by a test and reference illumination SPD. The MI_{UV} value provides an indirect measure of the relative UV content of the source, which is important in many print viewing applications where fluorescent media are used and the total reflectance has a significant fluorescent component.

To ensure consistency in the computation of CRI, MI_{VIS} and MI_{UV} values, ISO 13644:2009 provides a detailed computational procedure. As with the CRI, MI_{VIS} and MI_{UV} values are relative to the reference illuminant CIE D50.

The 2000-lux illuminance specified in ISO 3664:2009 P1 viewing condition corresponds to an adopted white luminance of 637 cd m^{-2}, which compares with a typical display luminance of 80-120 cd m^{-2} and an adopted white luminance in a more typical 300-500 lux viewing environment of 95-160 cd m^{-2}. The higher level needs additional evidence to justify it given the propensity for radiative damage and the greater energy consumption associated with it.

4.2. BS 5454:2000

BS 5454:2000 is primarily concerned with media of long-term significance. It recommends a maximum of 200 lux for display of materials with lower light sensitivity such as carbon black inks, and a lower 50 lux level for light-sensitive materials. It recommends a total exposure of 18,000 lux (18Klux) hours, and states that illumination should have a maximum of 10 watts of UV radiation per lumen of visible light. For viewing while in storage (for example for conservation purposes) an illuminance of 100-300 lux is recommended. Exposure to UV is limited to a maximum of 10 watts in the UV region per lumen of visible light.

In setting 18Klux total exposure complete reciprocity is assumed, as it is stated that "exposure of a document to 50lux for 10hrs is likely to cause the same deterioration as exposure to 500lux for one hour". However, reciprocity failure in radiative damage has been shown to exist [9-12].

4.3. ISO 18937

ISO 18937 is still under development. Its scope is to specify methods and procedures for measuring the indoor light stability of colour photographs. It specifies a range of test illumination conditions, sources, and associated longpass filters to control UV content of the illumination.

The viewing conditions, and associated illumination intensities, currently included are:

- indoor daylight (30-80klux)
- direct sunlight in-window (50-100klux)
- cool white fluorescent (30-80klux)
- other sources i.e. compact fluorescent, tungsten, LED, OLED, metal halide and others (20-80klux)

It is noted in ISO 18937 that reciprocity failures may occur in high-intensity light fading tests, and that "the closer accelerated test conditions are to actual use conditions, the more meaningful the test results become".

ISO 18937 also specifies humidity and temperature control for accelerated exposure tests, which are important as it has been shown that their effects can be significant [13].

5. Radiative damage to reflective media

Optical radiation induces a colour change in printed inks and papers. Such changes are usually in the direction of reduced contrast and colourfulness, and are therefore almost always undesirable. Some degree of change, such as a slight yellowing of the paper, is inevitable as documents age, but the goal of the conservator is to minimise such damage and preserve the appearance of the object with as much fidelity to the original as possible.

Oxidative changes to inks, quenching of fluorescent whitening agents and other changes are distinctly more likely with UV radiation than visible light, and since shorter wavelengths are more energetic it is often assumed that the degree of damage is a continuous function of wavelength. This assumption is broadly supported by experimental results [14, 15], but there is insufficient data on the damage-activation spectra of dyes, pigments and other materials to rely on the assumption and there remains the possibility that some materials may be selectively affected by wavelength.

Radiation intensity levels for viewing of photographs and prints is normally measured and specified in terms of photopic illuminance, whereby the irradiance is weighted by the CIE photopic luminous efficiency function over the range 360-830nm. The units of illuminance are lux. A measure in lux therefore correlates with the amount of light energy available to the human visual system, but not to the total radiative energy, since photopic luminous efficiency has a Gaussian-like shape that peaks at 555nm and asymptotes to zero at the red and blue ends of the visible spectrum.

6. Spectral power, correlated colour temperature and propensity for radiative damage

Figure 1 shows the spectral power distribution (SPD) of the radiant exitance of black body (or Planckian) radiators at four selected correlated colour temperatures, from 3000 to 6000 K. At 3000 K the SPD is very similar to that of unfiltered tungsten, while at higher CCTs the SPDs approximate that of natural daylight.

Journal of Physics: Conference Series **231** (2010) 012016 doi:10.1088/1742-6596/231/1/012016

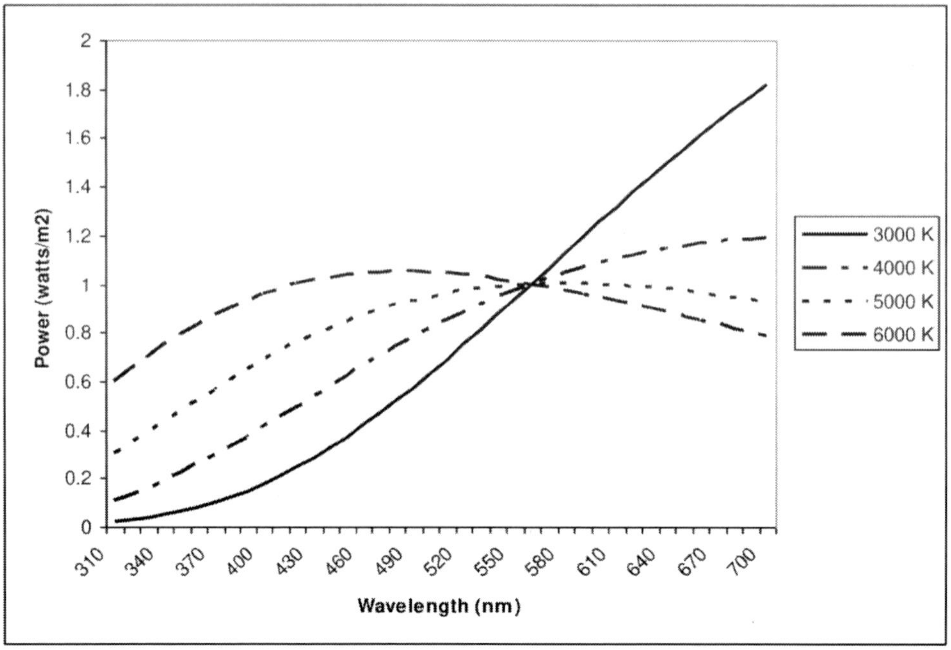

Figure 1. Radiant exitance of black body radiators at four different correlated colour temperatures

According to ISO 3664:2009, 'The human visual system is excellent at adapting to the colour temperature of illumination sources which approximate a Planckian radiator between 3000K and 6000K.' However, at low power such adaptation may not be complete and for example a source with a correlated colour temperature (CCT) of 3000K may appear yellowish, while a CCT of 6000K may appear bluish [16]. Daylight simulators with CCTs of 5000-6500K are in widespread use and are standardised in many industries such as paint, textile and papermaking. The CIE Colour Rendering Index defines the agreement between the spectral power distribution of a source and a reference illuminant of a similar CCT, but the illuminant used as reference can be any published illuminant; for example a low-power tungsten lamp will have a high CRI against a reference Illuminant A, but may lead to relatively poor visual discrimination of blue and green colours.

In a black body radiator, at higher correlated colour temperatures (CCT) the relative power at short wavelengths increases, as does the luminous efficacy. Hence the balance is between greater visual performance and energy efficiency against the likelihood that a higher proportion of more damaging short-wavelength radiation will be present.

While ISO 3664:2009 is primarily concerned with graphic arts viewing, Informative Annex C suggests that a CCT as low as 2800K and an illuminance as low as 375 lux are acceptable for exhibiting photographic prints.

Table 1 gives the illuminance (in lux), radiant exitance (in watts per square metre) and photon count (x 10^{18}) for the four black body radiators in Figure 1, in three wavelength ranges: the visible, the UV-A region and a short-wavelength region of the visible spectrum.

Table 1. Properties of black body radiators (normalised to 1.0 watt m^{-2} at 560nm)

CCT (Kelvins)	3000	4000	5000	6000
Photopic illuminance of black body radiator (lux)	7315	7185	7192	7231
Radiant exitance of black body radiator over visible spectrum 380-700nm (watts m^{-2})	29.93	28.85	30.00	31.72
Radiant exitance of black body radiator over short-wavelength region of visible spectrum 380-450nm (watts m^{-2})	1.86	3.80	5.87	7.86
Radiant exitance of black body radiator over UV-A region 310-400nm (watts m^{-2})	0.81	2.47	4.97	7.80
Photons sec^{-1} over visible spectrum (x 10^{18})	89.84	82.61	82.90	85.36
Photons sec^{-1} over short-wavelength visible (x 10^{18})	3.97	8.02	12.34	16.46
Photons per second over visible + UV-A (x 10^{18})	90.49	84.98	88.15	94.33

It can be seen that the black body radiators are normalised in such a way that the illuminance, radiant exitance and photon count values are very similar for all four CCTs. However, it can be seen from Figure 1 that the four radiators have very different relative power levels in the short-wavelength (blue) end of the visible spectrum. Clearly the black body radiators with higher CCTs have significantly more power in the shorter wavelengths.

Unless there is a wavelength-selective damage spectrum, it is likely that radiative damage correlates with the number of photons incident on the surface. The photon count in Table 1 was determined for the SPDs shown in Figure 1 from Planck's equation $E = h\nu$ or $E = hc/\lambda$ where c is the speed of light, λ is its wavelength and h is Planck's constant.

Although the propensity for damage continues to increase at shorter wavelengths than UV-A, this can be discounted for photographic display since oxygen in the atmosphere efficiently blocks the lower wavelengths, and glass used in windows and the envelopes around light sources is also normally opaque to the shorter wavelengths.

Table 1 suggests that the lux measure is likely to be a good predictor of the propensity for radiative damage for those sources which are similar to black body radiators and are filtered to remove wavelengths below 380nm. However, other sources, especially those with a higher proportion of short-wavelength energy, are likely to have a different and possibly higher propensity to cause radiative damage. For example, fluorescent lamps have a sharp peak in the relative spectral power distribution at 437nm and 'white' LEDs emit strongly in the blue region, at a peak wavelength which depends on the type of LED used.

7. Conclusions

Existing standards for the illumination of digital prints and photographs have focused on either the maximising visual performance or on limiting radiative damage, and for most applications there is a need to achieve a balance between these.

Recent work on the standardisation of conditions for the display of photographs and prints improve the definition of illumination sources for the goals of maximising visual performance and minimising radiative damage. However, certain assumptions are made in these specifications which require further evaluation. In particular, further work is needed to understand:

- the radiative damage spectrum for the materials used in digital prints
- the illumination levels and spectral power distribution which lead to optimum visual performance

Greater specificity is needed in the available data so that it is possible to select optimum illumination levels and spectral power distributions for particular applications. The requirement to maximise the energy efficiency of illumination also imposes a requirement to understand the minimum illumination intensities required for a given visual task.

Given that a simulator of daylight illumination is likely to give better performance on visual discrimination tasks, light source quality should in the absence of further data be defined in terms of the CRI and MI values relative to a reference D50 illuminant.

The propensity for radiative damage varies over the electromagnetic spectrum and for many sources the conventional lux measure may not adequately express this. Hence the use of watts m^{-2} is recommended in preference.

References

[1] Stevens S S 1957 On the psychophysical law *Psychological Review* 64(3):153–181
[2] ISO TC 130/WG 3 N0846 Softproof proposal (Committee document) ISO, Geneva
[3] CIE Publication 13.3 1995 Method of Measuring and Specifying Colour Rendering Properties of Light Sources, CIE, Vienna
[4] Manufacturer's data sheet for Philips TL950 lamp
[5] Murray M 2010 Cree Breaks 200 Lumen Per Watt Efficacy Barrier, Cree press release, http://www.cree.com/press/press_detail.asp?i=1265232091259
[6] ISO 3664:2009 Graphic technology and photography — Viewing conditions, ISO, Geneva
[7] BS 5454:2000 Recommendations for the storage and exhibition of archival documents, ISO, Geneva
[8] ISO 18937:2010 (WD) Imaging materials — Photographic reflection prints — Methods for measuring indoor light stability, ISO, Geneva
[9] Bush A Lindstrom B LaBarca J Bugner D E, and Kapecki J 2006 Rf method of reciprocity performance testing – A statistical approach to reducing the time needed to reach accurate conclusions – *Part 1, Proc Int. Conf. on Imaging Sci.* pp. 410-413
[10] Bush A Lindstrom B LaBarca J Bugner D E and Kapecki J 2007 Rf method of reciprocity performance testing – A statistical approach to reducing the time needed to reach accurate conclusions – *Part 2, Proc. IS&T NIP23: Int. Conf. on Digital Print. Tech.* p733
[11] Bugner D and Suminski C 2000 Filtration and Reciprocity Effects on the Fade Rate of Inkjet Photographic Prints, Proc. *IS&T NIP16: Int. Conf. on Digital Print. Tech.*, pp. 90–94
[12] Guo S and Miller N 2001 Estimating Light-Fastness of Inkjet Images: Accounting for Reciprocity Failures, *Proc. IS&T NIP17: Int. Conf. on Digital Print. Tech.*, pp. 186–191
[13] Reber J and Hofmann R 2005 Humidity Effects on Light Fastness Testing, *Proc. IS&T NIP21: Int. Conf. on Digital Print. Tech.* p344-347
[14] Baumann E and Hofmann R 2004 The Equivalence of Light Sources in Light Stability Testing, Proc. *IS&T NIP20: Int. Conf. on Digital Print. Tech.*, pp. 699–703
[15] Saunders D and Kirby J 1994 Wavelength-dependent fading of artists' pigments, in Preventive Conservation: Practice, Theory and Research, A. Roy, and P. Smith (eds.), (London, Int. Inst. for Conservation), pp. 190–194
[16] Davis R Ginthner D 1990 Correlated color temperature, illuminance level, and the Kruithof curve, *J. Illuminating Eng. Soc.* **19** pp. 27–38

An investigation into the effects of solvent content on the image quality and stability of ink jet digital prints under varied storage conditions

A L Fricker[1], **A Hodgson**[2] **& M Sandy**[1]

[1] MATAR Research Centre, University of the Arts London, London, UK.

[2] 3M Security Printing and Systems Limited, Gorse Street, Chadderton, Oldham, UK. OL9 9QH

a.fricker@lcc.arts.ac.uk

Abstract. Increasing numbers of galleries, museums and archives are including ink jet printed materials into their collections, and therefore displays. There is evidence that the instability of these prints is such that images can suffer deterioration in print quality or in extreme cases, a loss of information over an extended period of time. This is shorter than the period typically required for perceptible deterioration to occur in many other paper-based artworks. The image stability of prints is affected by a number of factors some of which have already been studied. However the role played by the ink solvent in the loss of image quality has yet to be explored. This paper will outline research being undertaken to investigate the effects of solvent content which may increase/promote the loss in image quality of the hard copy prints when stored or displayed under a range of temperature and humidity conditions.

1. Introduction

Ink jet digital printing is now well established as a fine art and photographic medium. These images are found in many museums, gallery collections and archives, where longevity of collections is an important issue.

Consumer expectation of image stability for ink jet prints is mistakenly high, due to the standards set by traditional silver halide based colour photography in particular dye destruction processes. The digital print industry is more motivated by production rather than preservation of its products and it is the heritage groups who are most concerned with longevity issues. Importantly, the shortfall in stability of ink jet inks is not generally appreciated by the consumer public who are often unprepared for longer term disappointment.

The image stability of ink jet prints is affected by four primary factors: light, atmospheric pollutants, heat and humidity; in addition, fungal growth and physical damage will also cause losses in image quality.

The roles of light and atmospheric pollutants have been explored and their effects on print stability extensively studied [1, 2]. As a result, procedures for the control of these factors are in an advanced state of preparation [3, 4].

It has been observed that variations in heat and humidity also cause loss of clarity and degradation of image quality in ink jet prints as a result of dye and/or pigment migration in the substrate. The reasons for degradation effects caused by variations in temperature and relative humidity have not been fully established and consequently, conservation procedures have not been standardised; at least three approaches are currently in use [5, 6, 7]. Furthermore, the role of residual solvent in the dye/pigment migration process has not been investigated. This paper will outline the ongoing research being performed by the authors in this area.

2. Ink Jet Inks

Ink jet inks consist of a number of different chemicals designed to improve the printer performance and the quality of the final print. The majority of commercial inkjet printers used for fine art production contain aqueous inks, where water is the primary solvent and accounts for up to 95% (by weight) of the ink. The remainder of the ink will typically consist of the colorant, humectants, surfactant and other additives [8].

Colour in ink jet prints results from the chromophores on either dye or pigment molecules. Optically, dyes scatter little light and therefore provide high saturated colour. They are, however, vulnerable to degradation via UV light and atmospheric pollutants and are relatively sensitive to humidity. Pigment based inks are usually more durable and resistant to environmental factors. However, the relatively large pigment size results in increased surface roughness of the ink layer and causes light to scatter, thereby producing less saturated and duller prints. Pigment particles can also cause clogging of the nozzles in ink jet printers, reducing performance and print quality.

Penetrants are added to the ink to promote penetration of the ink into the paper. These co-solvents also provide shorter drying times for the ink and many also act as humectants, reducing water loss from the ink, so that it does not dry out or clog printhead nozzles. Smooth jetting and penetration of the ink is also aided by the presence of surfactants, which adjust the surface tension of the ink.

Additional components of the ink perform a variety of functions. These include buffers to maintain pH levels, chelating agents to reduce print head corrosion and additives to reduce cockle and curl of the paper fibres.

3. Print Media and Substrate Considerations

The ink-substrate interactions are arguably the most important mechanisms governing print quality. The print characteristics directly resulting from these interactions include optical density, dot size, dot gain, colour, resolution and edge sharpness. These are highly dependent on drop spreading, which is governed by the viscosity and dynamic surface tension of the ink as well as the absorption properties of the substrate. An example of different printed dot morphologies is illustrated in figure 1. [9]

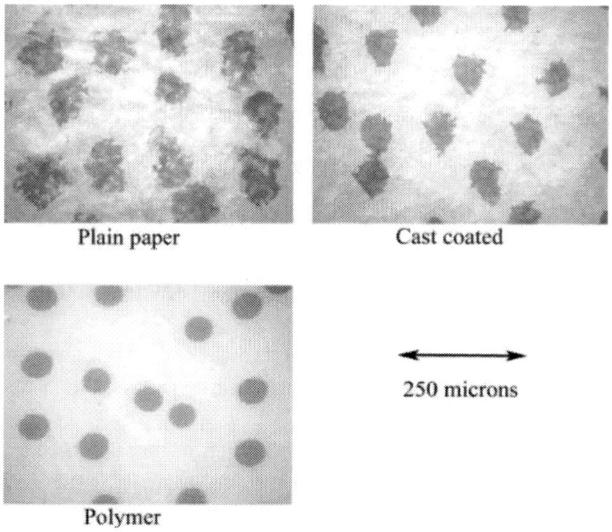

Figure 1. Dot morphologies on different inkjet media

A range of papers can be used with ink jet inks and good printability depends on a number of factors [8, 10]. Plain papers provide the lowest image quality. Due to their high absorbency of water, large areas of print can cause the substrate to cockle and wave. Some spreading of the ink also occurs, resulting in duller and less sharp images. As a result higher performance (and value added) products have appeared such as fine art papers and products with swellable or porous coatings.

Much sharper images are produced by using swellable or porous coated products. Swellable papers can only be used with dye based inks. They work by having a coating which swells to absorb the printed dye. The disadvantage of these papers is that it can take several hours for prints to dry. Microporous papers, on the other hand, have small pores in their coating which rapidly draw water further into the paper, resulting in a print dry to the touch. These papers can be used with both pigments and dyes. SEM micrographs of plain and microporous papers are shown in figures 2 and 3.

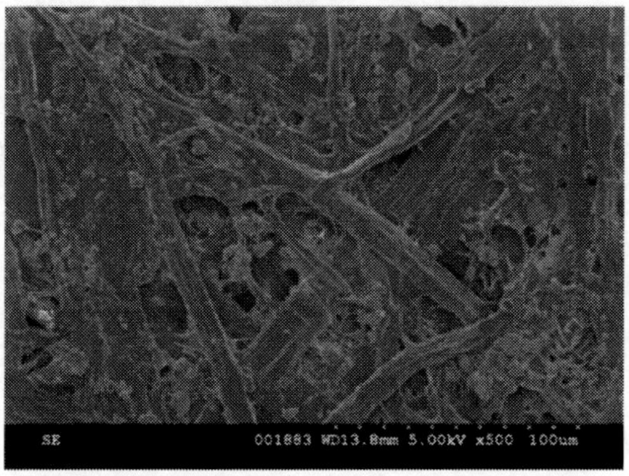

Figure 2. SEM image of plain paper

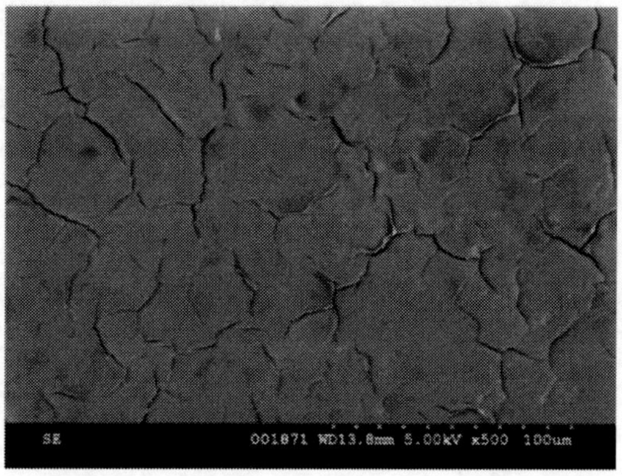

Figure 3. SEM image of coated microporous paper

Finally, fine art papers are made of 100% cotton rag. Image quality is similar to that observed for plain papers. Some manufacturers have coated the papers with a porous ink jet receiver layer to improve the image quality. These papers do not include a resin coated layer.

4. Print Drying and the Role of Solvent

It is believed that it takes a maximum of 8 hours for the water content in an inkjet print to evaporate fully [11]. Although microporous papers appear dry immediately after printing (due to the surface pores), these prints also take time to reach moisture equilibrium with the surrounding atmosphere. Consequently, it has been noted that there is the potential for short term colour drift in the prints in the hours after printing [12]. This indicates the need for conditioning procedures when considering image permanence test methods.

Although the carrier fluid in aqueous inks is principally water, these inks also contain co-solvents. These solvents act as humectants and penetrants, reducing water loss and promoting penetration of ink into the paper. The presence of these co-solvents also impacts the evaporation and drying time of the ink.

While the majority of solvent will evaporate from a print within several days, small quantities of solvent remain in the print for a longer period of time. It is thought that this residual solvent causes the ink dyes and pigments to migrate further over periods of months and years, resulting in loss of image quality and definition. It is also likely that this migration is affected by temperature, humidity and other storage conditions.

5. Background and Previous Work

It is known that both lateral and vertical diffusion of inkjet colourants occurs under variable conditions of temperature and humidity [6, 7, 12-16]. Lateral diffusion results in loss of image clarity due to uneven, uncontrolled spread of ink across the printed sheet; this is sometimes described as line quality retention. Vertical diffusion of ink into the body of the substrate results in numerous variations to image quality. These include a lowering of image density and gloss as well as loss of colour and variations in colour balance. Additionally, as the different colour process inks exhibit different diffusion properties, undesirable hue shifts can occur due to the uneven migration of the different inks.

Numerous authors have investigated the effects of humidity and temperature on the final print [17, 18, 19]. Dye based prints have shown increased susceptibility to changes in humidity, particularly when printed on a swellable medium [7, 15]. In contrast, pigment based inks appear to be relatively stable although studies have shown that they can demonstrate migration under conditions of high humidity [20, 21]. Substrate properties also affect the permanence of the final print. Research has shown that the fluorescence of whitening agents in the paper is affected by temperature but not significantly by humidity [22].

These factors all have adverse effects on the ability to maintain a stable, consistent hard copy and present problems for museums and galleries with the storage and display of works of art produced using ink jet inks. In addition, digital ink formulations involve the inclusion of high boiling point solvents such as glycols which improve ink mobility within the print head but further complicate the diffusion processes in the printed hard copy.

Research to date has only studied the consequences of these changes and has not attempted to explore what is happening within the printed system. It is envisaged that effect of solvents and dye/pigment properties together with the structure of the substrates and changes in temperature and humidity will all play a part in affecting the degree of dye/pigment migration.

Previous studies by Steiger and Brugger [23] found that residual solvents in dyes can decrease their fastness behaviour. This is thought to be because of a reduction in dye aggregation resulting from continued diffusion of the solvent through the substrate. Research by Wilhelm [11] has investigated the lightfastness of ink jet prints under differing drying conditions. It was shown that the light fading of prints can display very different behaviour depending on the drying conditions involved. In the case of the magenta dye, contact drying of prints results in better light fastness than air drying. This is possibly a result of the contact drying removing the residual solvent content in the print, suggesting that these reduce dye aggregation and therefore lightfastness.

This research will examine the effects of temperature and humidity on the rates of diffusion that occur for different process colour inks on a range of substrate types. The role of the solvent will be closely examined because this will play a major part in dye diffusion. It is hypothesised that the solvent will continue to diffuse through the paper after drying, resulting in the migration of ink dyes and pigments further into the substrate over time. It is also hypothesised that lateral and vertical migration of these pigments and dyes could, with time, result in a decrease in line quality and sharpness as well as colour and balance, affecting the quality of the image as a whole. The controlled removal of solvent will be explored as a means to stabilise prints under all conditions of exposure to relative humidity and heat, rather than simply storing them under standard conditions where deterioration will still occur.

6. Experimental Method

In order to conduct tests for differing levels of humidity fastness, a range of inks and substrates will be used to evaluate the effect of solvent content. These will include both dyes and pigment based inks as well as swellable, microporous and fibrous papers. Ink jet prints will be obtained from commercial printers who are able to reliably print and store under controlled conditions of temperature and relative humidity. These prints will be allowed to condition at constant temperature and relative humidity before being subjected to test methods.

The test images will be selected to objectively measure colour changes and loss in line quality. The printed image will include two picture elements to enable comparative visual assessment coupled with a target. The target will include a half-tone chequer-board pattern of colours similar to that used by Kaimoto and Shibahara [13] to record the edge definition and image area. These measurements will be made using optical microscopy coupled with image analysis. Changes in the density, gloss and hue of each colour patch will be recorded in order to monitor colorant migration. Solid colour targets will also be included to enable the extent of vertical ink diffusion to be recorded using scanning electron microscopy. All combinations of the four process inks will be studied to assess the effects of interactions between them.

The relative solvent content of the samples will be initially measured using a variety of techniques to establish the most effective. The techniques include Fourier-transform infra red spectroscopy (FTIR), thermal analysis techniques of thermogravimetric analysis (TGA) and differential scanning calorimetry (DSC) and gas chromatography.

Following analysis, half the samples will be sealed in impermeable enclosures to ensure no loss of solvent to the storage atmosphere. One third of the sealed and unsealed samples will be stored under known temperature and relative humidity conditions as recommended in BS5454. These samples will constitute the control groups. A second third will be further subdivided and each sample stored at different temperatures and relative humidity conditions ranging from 20°C to 80°C and 20% to 80% RH; this will also enable an assessment of the impermeability of the enclosures. The final third of samples will be subjected to a range of treatments to remove or reduce the ink solvent content before storage under the same conditions. Methods to remove residual ink solvent from the prints will include the use of a degasser and absorbent contact drying of prints. The reduction in solvent content will be monitored using FTIR, thermal analysis and headspace analysis gas chromatography.

Stored samples will be measured at regular intervals up to a total storage time of two years. Changes over this period are known to occur from light fading studies [7] where changes have been detected over 24 months. Measurement techniques will be densitometry and glossmeter for changes in image density and gloss, and colorimetry for hue shift. Changes from the initial measurements will be plotted on logarithmic scales and analysed using the well established Arrhenius method [24]. The picture elements will permit subjective assessment of perceptible changes in image quality by a panel of observers.

Correlations will be sought between residual solvent content and changes in colorant migration, density and hue shift. If such a link is established, procedures for long term preservation of prints will follow.

7. Conclusion

Research findings from the project will provide information about the role of residual solvent in the migration of ink in papers over long periods of time. The findings will benefit two audiences, those producing ink jet prints and those responsible for the safe keeping of such prints. The first group comprising artists, commercial printers and domestic users who will be informed of the published protocols and guidelines to ensure the minimum deterioration of their ink jet prints. The second group, comprising conservators and curators will have recommendations and guidance on how best to store and display artefacts containing ink jet prints.

Acknowledgements

The authors would like to acknowledge the support of the Arts and Humanities Research Council (AHRC) for providing the funding for this project.

References

[1] Vogt B and Frey F 2001 Issues in evaluation and standardisation of light fading tests of ink jet materials *IS&T's NIP 17th Int. Conf. on Digital Printing Technologies* pp 218-221

[2] Saitmacher K et al. 2003 Image stability of ink jet prints. Investigations on the impact of ozone *IS&T's DPP Int. Conf. on Digital Printing Technologies* pp 201-202

[3] ISO 18937: Imaging materials - Photographic images - Methods for measuring indoor light stability

[4] ISO 18941: Imaging materials - Colour reflection prints - Test method for ozone gas fading

stability

[5] Adelstein P Z 2006 International standards on permanence of digital information *Proc. 3rd Int. Conf. on Preservation and Conservation Issues related to Digital Printing and Digital Photography* pp 36-39

[6] Bugner D 2006 An update on the effect of arrhenius testing of ink jet photographic prints *Proc. 3rd Int. Conf. on Preservation and Conservation Issues related to Digital Printing and Digital Photography* pp 108-113

[7] McCormack-Goodhart M and Wilhelm H 2005 New test methods for evaluating the humidity fastness of inkjet prints *Proc. 'Japan Hardcopy 2005' The Annual Conf. of the Imaging Society of Japan* pp 95-98

[8] Jürgens M 1999 Preservation of ink jet hardcopies *Capstone Project, Cross-Disciplinary Studies, Rochester Institute of Technology*

[9] Hodgson A and Jackson A M 2004 The light fading of dye based inkjet images - a multidimensional issue *IS&T's Archiving* pp 43-48

[10] Barba i Mundó R 2003 The development of inkjet printing media for digital production printing *IS&T's DPP2003: Int. Conf. on Digital Production Printing and Industrial Applications* pp 218-219

[11] Wilhelm H 2004 How long will they last? An overview of the light-fading stability of inkjet prints and traditional color photographs *IS&T's 12th Int. Symp. on Photofinishing Technology* pp 32-37

[12] McCormick-Goodhart M and Wilhelm H 2001 The influence of relative humidity on short-term color drift in inkjet prints *IS&T's NIP17: Int. Conf. on Digital Printing Technologies* pp 179-185

[13] Kaimoto H and Shibahara Y 2007 Test methods of humidity fastness of inkjet printing materials *IS&T's NIP23: Int. Conf. on Digital Printing Technologies* pp 728-731

[14] Hill P Suitor K and Artz P 2000 Measurement of humidity effects on the dark keeping properties of inkjet photographic prints *IS&T's NIP16: 2000 Int. Conf. on Digital Printing Technologies* pp 70-73

[15] Bugner D E and Lindstrom B L 2005 A closer look at the effects of temperature and humidity on inkjet photographic prints *IS&T's 21st Int. Conf. on Digital Printing Technologies* pp 348-352

[16] Onishi H Hanmura M Kanada H and Kaieda T 2001 Image permanence of ink jet photographic prints *IS&T's NIP17: 17th Int. Conf. on Digital Printing Technologies* pp 192-196

[17] McCormick-Goodhart M and Wilhelm H 2000 Humidity-induced color changes and ink migration effects in inkjet photographs in real-world environmental conditions *IS&T's NIP16: 2000 Int. Conf. on Digital Printing Technologies* pp 74-77

[18] Högberg O Talaskivi M and Ström G 2001 Humidity effects on plain paper in inkjet printing *IS&T's NIP17: 17th Int. Conf. on Digital Printing Technologies* pp 874-877

[19] Oakland M Bugner D E Levesque P and Artz P 2001 Arrhenius testing under controlled atmospheres: measurement of dark keeping properties of inkjet photographic prints *IS&T's NIP17: 17th Int. Conf. on Digital Printing Technologies* pp 167-170

[20] McCormick-Goodhart M and Wilhelm H 2003 The correlation of line quality degradation with color changes in inkjet prints exposed to high relative humidity *IS&T's NIP19: 2003 Int. Conf. on Digital Printing Technologies* pp 420-425

[21] Work R A and Brown R T 2001 Dye and pigment ink jet image stability and permanence *IS&T's NIP17: 17th Int. Conf. on Digital Printing Technologies* pp 414-417

[22] Blayo A and Medlege F 2001 Aging properties on non-impact prints: influence of the paper properties *IS&T's NIP17: 17th Int. Conf. on Digital Printing Technologies* pp 235-238

[23] Steiger R and Brugger P A 1998 Photochemical Studies on the Light Fastness of Inkjet Systems *IS&T's NIP13: 13th Int. Conf. on Digital Printing Technologies* p 114

[24] ISO 18924:2000(E): Imaging materials - Test method for Arrhenius-type predictions

Journal of Physics: Conference Series **231** (2010) 012018 doi:10.1088/1742-6596/231/1/012018

The influence of the accelerated ageing on the black screen element of the Electroink prints

I Majnaric, Z Bolanca and I Bolanca Mirkovic

University of Zagreb Faculty of Graphic Arts, Getaldiceva 2, 10 000 Zagreb, Croatia

E-mail: majnaric@grf.hr; zbolanca@grf.hr; ibolanca@grf.hr

Abstract. Printing material and prints undergo changes during ageing which can be recognized in deterioration in the physical, chemical and optical properties. The aim of this work is to determine the optical changes of the prints caused by ageing of the printing material and of the prints obtained by the application of the indirect electrophotography. The change of the screen elements in lighter halftone areas, which was obtained by the usage of the microscopic image analysis, has been discussed in the article. For the preparation of samples the following papers were used: fine art paper, recycled paper and offset paper as well as black Electroink. Three sample series were observed: prints on nonaged paper and ElectroInk, prints on aged paper and ElectroInk and prints on aged paper and nonaged ElectroInk. The investigation results show that by ageing of the uncoated printing substrates the decrease of the dots on prints can be expected, while the printing on the aged paper results in the increased reproduction of the halftone dots. The obtained results are the contribution to the explanation of the influence of the accelerated ageing process of papers which are used for printing and the aged prints on the halftone dot changes. Except the mentioned determined scientific contribution the results are applicable in the area of the printing product quality as well as in the forensic science.

1. Introduction

Electrophotographic printing principle with liquid toner is done by means of the satellite printing unit with the organic photoconductor, and indirect printing with the intermediate blanket which enables 100% ink transfer on the printing substrate [1].

ElectroInk with the starting low dynamic viscosity coefficient rapidly changes during printing. Such viscosity change is enabled by the non polar isoparaffinic hydrocarbon with the evaporating temperature of 40°C. Because of the temperature of 125°C which is achieved in printing, Electroink ceases to be in the liquid state and it becomes the sticky pigment paste. By heating the pigment polymer, the chemical process of plasticization emerges. By the plasticization process the pigment polymer obtains the characteristic spongy shape which is optimal for adhering to the printing substrate. The final print is characterized by a thin ink layer (about 1 μm) which, by its chromaticity and saturation, is the nearest to the print made by the classical offset printing process [2].

In the basic physical state, the black Electroink is the emulsion containing the coloured thermoplastic polymer (about 30%), which is dispersed in the non polar easily volatile oil (about 75%). Except that it contains the agents for increasing the electric conductivity - negative charge director (about 1%). In the thermoplastic polymer, the particles of the black pigment are built in, the share of which varies depending on chemical composition (pigment particles of the type Black 7 are contained from 10 to 35% while the ink carriers based on dye are contained from 3-25%). In the final phase, the ferromagnetic materials (such as iron, nickel and cobalt) can be added, the optimal share of which is 65%. The particle size of the black pigment is between 0,1 and 5 μm. The particles of the thermoplastic toner have a characteristic star like shape. Such shape serves primarily for better directing of the Electroink movement, i.e. for easier accepting the additives for the regulation of electric conductivity. They are explicitly polar molecules and catch the star like pigments with their more positive side. The caught pigment particles are now ready for directed movement toward the less negative electrostatic field [3].

The used black pigment based on carbon (CI Pigment Black 7) is one of the most important black ink carriers. Such pigments are extremely tiny particles, the inking density of which depends on the production method. Non rubber varieties, used in printing inks are produced in pallet, bead, dense or fluffy form. They are chemically inert, extremely fast to heat, light, acids, alkali, solvents and soap. However they do contain a small proportion of volatile material, the effect of which may become apparent on dispersion: accounts for over 95% black ink usage [4].

Generally speaking, the printing inks are the mixture of different components which are significantly dependent on the printing technology. The selection of suitable inks for printing and investigation of ageing stability of ink film are scarce. It is published in a study of stability for offset printing inks on alkaline paper upon damp heat and light ageing [5].

Their results indicate that black ink keeps the chemical and visual properties upon accelerated ageing, lower ageing stability especially for magenta and yellow prints. Halinova and coauthors studied changes in optical density, visible and FT-IR spectra and total colour differences of CMYK offset ink films printed on alkaline paper during accelerated ageing procedures with the aim to decide on the combination of paper vs. ink for permanent offset printing [6].

Deterioration in quality of an aged paper or prints can manifest itself in the physical mechanical, chemical and optical properties. In regard to the changes of paper over time, the term durability is interesting. Paper durability depends on the physical mechanical properties of raw materials used in paper production (fibers, sizing agent, additives, fillers) as well as on environmental elements such as microclimatic conditions (heat, humidity, light), pollutants present in the air (ozone, oxides of sulphur and nitrogen), traces of transition metal ions (Fe (III), Cr(III), Cu(II)), biological agents (micro-organisms, mould) and printing processes (composition and kinds of inks) [7-10].

The permanence of paper depends on the chemical resistance of its components and on the influence of external factors. Paper permanence includes such things as lightfastness and refers to the ability of printing inks to resist fading or changing of colour upon exposure to light or weather [11,12].

Discolouration of a paper may be caused by the formation of chromophores upon ageing as a result of exposure to other light and volatile gases. Many volatile compounds as well as alcohols, ketones, aldehydes, carboxylic acids aromatic and aliphatic hydrocarbons and ethers can be released from paper during degradation processes depending upon paper chemical compositions [13].
The investigation results of the screen dots change on the prints obtained by indirect electrophotography using the accelerated aged and non aged printing substrates for the reproduction in regard to the accelerated aged prints are presented in this work. The obtained results are the

contribution to the explanation of the influence of the ageing process of the printing materials in the function of the reproduction characteristics in regard to the accelerated aged print. Except the mentioned contribution, the results are applicable in the area of the printing product quality as well as in forensic science.

2. Experimental

For the production of samples the electrophotographic printing machine HP Indigo Series 1000 was used. The printing form contains the following elements: multi-colour halftone photographs, patches for the determination of the full tones inking density, patches for the determination of the halftone value, patches for the determination of the relative printing contrast, patches for the determination of trapping, millimeter scales for the determination of the dimensional stability of prints, marks for the register control, achromatic areas for the determination of the grey balance and the standard FOGRA PMS wedges.

The uncoated natural paper, glossy fine art paper and the recycled paper were used as the printing substrates.

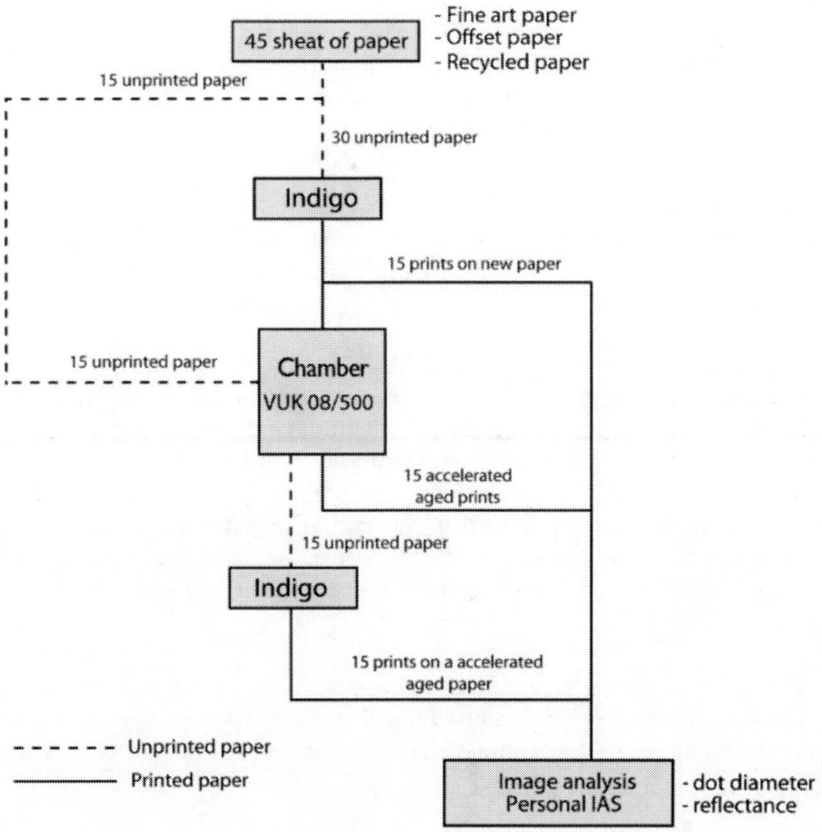

Figure 1 Experimental flow

For obtaining the measurable samples of different ages, paper and prints were aged in several phases:

1. phase: printing of non aged papers (2/3 of total printed run)
2. phase: accelerated ageing of the non printed papers (1/3 of samples) and ageing of 1/3 previously printed samples (made in the phase 1).
3. phase: printing of 1/3 of aged unprinted paper

As it is visible from the presentation (figure 1), three real printing interactions were obtained: non aged paper – non aged ink, aged paper – aged ink and aged paper- non aged ink. The ageing of samples is in accordance with the standard ISO 5360 [14]. The following controlled conditions were used: air temperature of 80°C, relative humidity of 65% and the ageing period of 24 days, without the influence of radiation. The process of accelerated ageing was performed in the microclimatic chamber VUK 08/500 [15].

The diameters of the smallest achromatic printing elements (prints in the black halftone area from 10%, 20% and 30%) were analyzed with the equipment for image analysis (QUE Personal IAS). Before the performed image analysis the samples were stored in accordance with the TAPPI standard T 1219sp-98 [16].

The detailed densitometric analysis was done on a very small halftone dot. The relation of ageing and the quality of the screen reproduction is presented by measuring the optical reflectance from the surface with only one black printed screen element.

3. Results and discussion

Diameter deviation of the halftone dots caused by the accelerated ageing for all three paper kinds (fine art paper, offset paper and the recycled paper) are presented in figure 2. The surfaces of 2,5 mm^2 within the area of 10%, 20% and 30% tone value were analyzed.

There is a general rule concerning the offset paper, that the aged prints form the smallest diameter of the screen elements, while the greatest halftone elements are obtained by printing on previously aged paper. One defective printing within the 20% halftone area (the smallest size of the printing elements was achieved for the non aged print) is noticed.

On the non-aged print on offset paper, in the area of 10% halftone value (160 elements) the average size of the screen elements $d_{10\%}$= 44,77 μm is noticed. With the ageing process, the screen elements in the same area are decreased from Δd=-14,05 μm, while the print on the aged offset paper gives the increase from Δd=0,18 μm.

For the area of 20% halftone value the non aged printed elements (245 elements) have the average diameter of $d_{20\%=}$ 63,69 μm. The standard deviation within the printed elements is very big SD=31,77 μm. With the ageing itself the halftone elements increase for Δd =1,54 μm, while the printing on the aged paper gives the increase of Δd=15,47 μm.

In the darkest investigated area (30%RTV-a) the non aged printed elements (224 elements) give the smallest average diameter $d_{30\%=}$ 98,02 μm. With the accelerated ageing the decrease of the screen elements of Δd =-11,89 μm is achieved, while the printing on an aged paper gives the increase of Δd =7,44 μm.

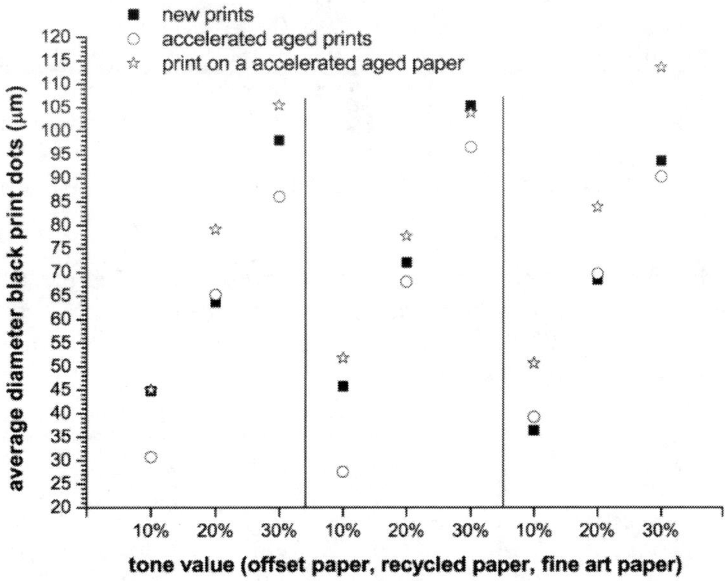

Figure 2 Diameter deviations of the halftone dots caused by the accelerated ageing

Prints made of the recycled paper are characterized by the regular change of the printed elements in relation to the halftone value. The increase of the halftone value for 10% results in the increase of the screen elements for approximately d_{med}=30,00 μm. On non aged print made on the recycled paper, in the area of 10% halftone value (177 elements) the average screen element size of $d_{10\%}$= 45,69 μm is noticed. The ageing process causes the decrease of the screen elements in the same area of Δd=-18,12 μm, while the print on aged offset paper gives the increase of Δd=6,11 μm. In the area with 20% halftone value the non aged printed elements (245 elements) give the average diameter of $d_{20\%=}$ 72,04 μm.

The standard deviation within the printed elements is great SD=31,77 μm. With the ageing itself, the screen elements decrease for Δd =-4,1 μm, while the printing on aged paper gives an increase of Δd=6,66 μm. In the area with 30% halftone value, the non aged printed elements (210 elements) give the smallest average diameter $d_{30\%=}$ 105,5 μm. With the accelerated ageing the decrease of the halftone elements of Δd =-8,89 μm is realized while the printing on aged paper ensures the decrease of Δd =-1,59 μm. The process of accelerated ageing (the aged print and the print on aged fine art paper) gives the greatest oscillations within the printed halftone elements. In this case the smallest halftone elements are formed.

On the non aged print made on fine art paper (gloss), in the area of 10% halftone value (177 elements) the average halftone size of $d_{10\%}$= 36,41 μm is noticed. In the ageing process the halftone elements in the same area increase for Δd= 2,79 μm, while the print made on aged offset paper gives the increase of Δd=14,23 μm. In the area with 20% halftone value the non aged printed elements (219 elements) give the average diameter of $d_{20\%=}$ 68,43 μm. The standard deviation within the printed elements is great SD=25,2 μm. With the ageing itself, the screen elements increase for Δd =-4,19 μm, while the printing on the aged paper gives an increase of Δd=15,47 μm. In the area with 30% halftone value, the non aged printed elements (217) give the smallest average diameter $d_{30\%=}$ 93,67 μm. With the accelerated ageing the decrease of the halftone elements of Δd =-3,4 μm is realized while the printing on aged paper ensures the increase of Δd =-19,89 μm.

On the non aged printed papers the standard deviation of the black halftone dots is SD_{off}=31,77μm, SD_{rec}=33,54 μm, SD_{gloss}=25,2 μm. With the accelerated ageing the deviation of the halftone dots is decreased (SD_{off}=22,36 μm, SD_{rec}=19,42 μm). The exception is the fine art paper, SD_{gloss}=28,68 μm. With the printing on the aged paper the results with the same tendency are achieved, i.e. the decreased deviation of the screen elements as with the aged print (SD_{off}=25,44 μm, SD_{rec}=24,27 μm, SD_{gloss}=33,94 μm).

The analysis of only one halftone dot (the characteristic dot in the area of 20% halftone value) is presented in figures 3,4,5. The densitometric reflectance (R%) on different aged prints is also presented. It is generally valid for all the presented curves that great reflectance from the reproduced dot gives the contrast picture and greater darkness.

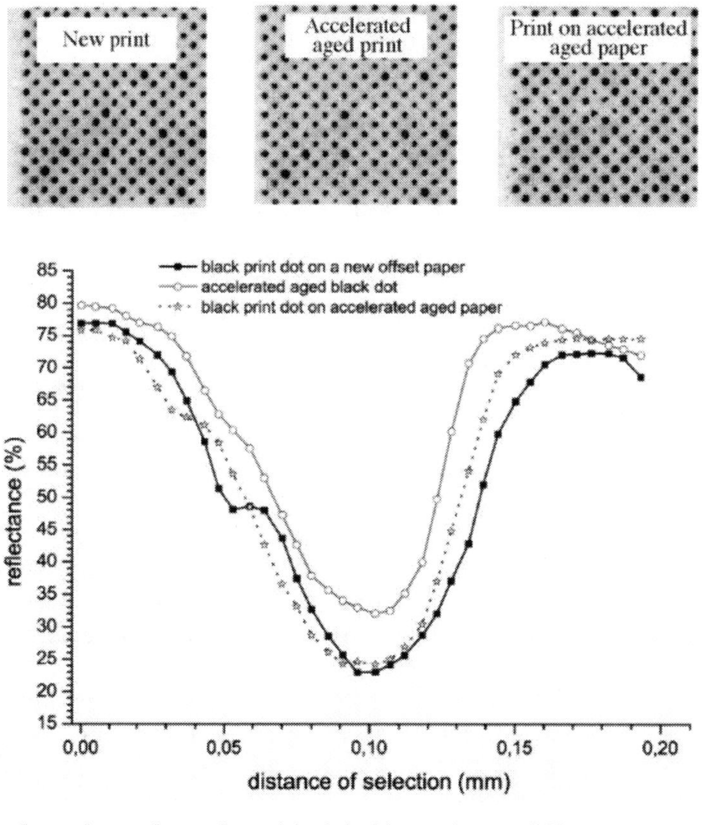

Figure 3 Reflectance from the surface of one black halftone dot on differently aged prints made on offset paper

With the image analysis (threshold 60%) of the surface from three halftone dots (selected area in the lower left angle) the following diameters are obtained: d=90,06 μm (non aged printed dot), d= 76,29 μm (accelerated aged dot), d=87,81 μm (non aged dot on the aged paper).The reproduced halftone element which best matches the ideal Gauss's curve is the aged offset print. Great steepness and the small reflectance with the maximum of R= 32,036% is characteristic for this curve. It is the direct consequence of ageing of the black pigment and the offset printing substrate. The other two curves are not so steep, which means that the dots have somewhat greater halo. The new printed halftone dot has the greatest reflectance (R=22,911 %), while the layer of the black Electroink on aged paper gives somewhat smaller reflectance (R=24,223%). Printing on the aged paper gives much broader peak inking density which can be attributed to the physical chemical ageing process of the non coated printing substrate.

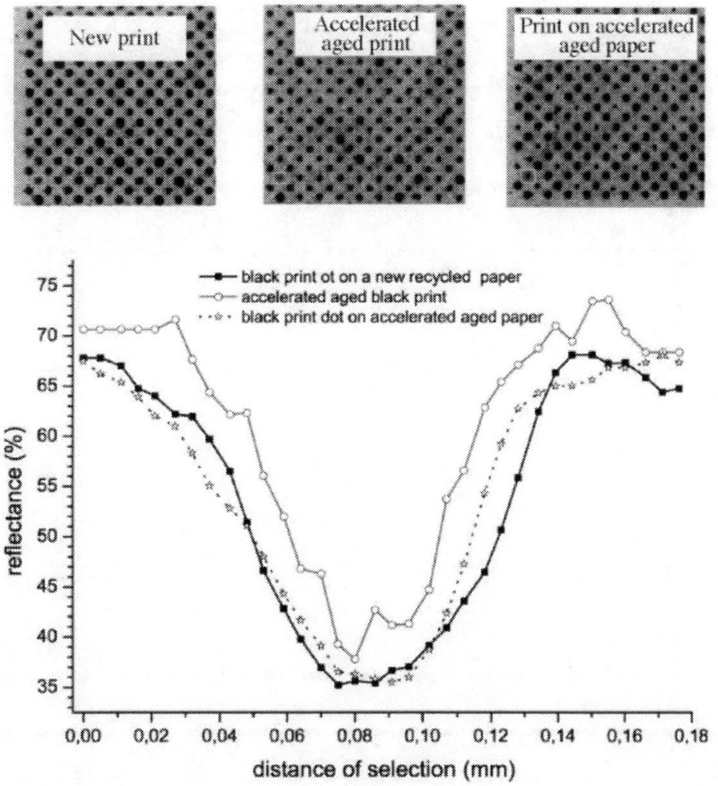

Figure 4 Reflectance from the surface of one black halftone dot on differently aged prints made on the recycled paper

With the image analysis of the printed recycled printing substrate (identical position) the following diameters are obtained: d=80,24 μm (non aged printed dot), d= 64,11 μm (accelerated aged dot), d=87,19 μm (non aged printed dot on aged paper).The least steep reflectance curve is obtained from the halftone dot surface which was printed after the accelerated ageing of the recycled paper. This resulted with great screen halo which is the direct result of the accelerated ageing of the printing substrate, which is visible yellow (R=35,497%). The non aged printed halftone dot has also high reflectance (R=35,221 %), which is held on greater part of the central dot (halo is not great).

The reproduced halftone element which matches the ideal Gauss's curve least of all is the aged print made on the recycled paper. High steepness and two peaks with the maximal reflectance R=37,778%

and R=41,133% are characteristic for that curve. With it, the worst reproduction of the black halftone element is achieved.

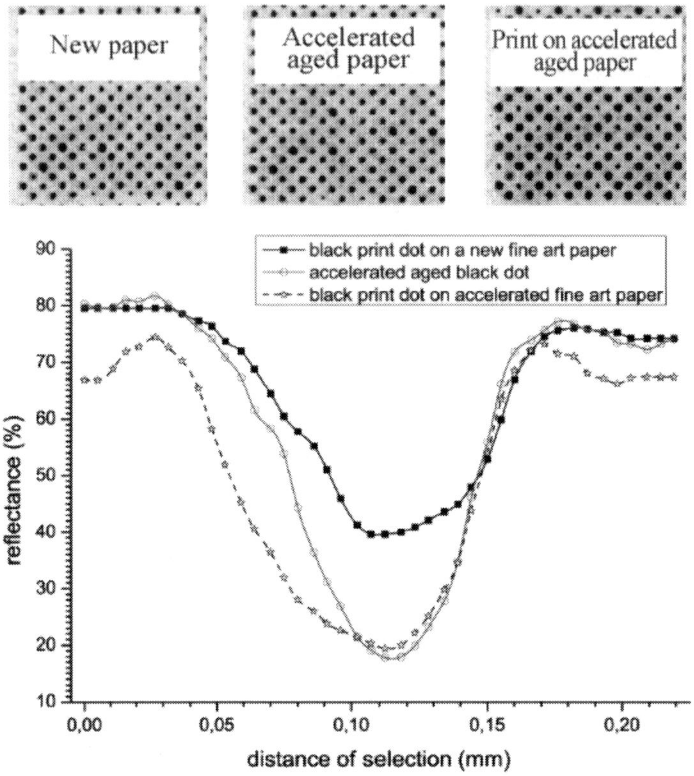

Figure 5 Reflectance from the surface of one black halftone dot on differently aged prints made on fine art paper

With the image analysis of the fine art paper of high quality, the following diameters are obtained: d=55,12 μm (non aged printed dot), d= 79,56 μm (accelerated aged dot), d=113,51μm (non aged printed dot on aged paper). In relation to the other analyzed printing substrates, it is characteristic, that the analyzed area on the non aged fine art paper has not got great reflectance (R_{max}=39,624%). Accelerated ageing influences the physical chemical change of the paper coating itself which influences the darkening of prints.

The reproduced halftone element which best follows the ideal Gauss's curve is the one which is submitted to the accelerated ageing. In this connection the maximal reflectance R_{max}= 17,778 % is achieved. Printing on the aged paper gives a very big halftone element in which the broad halo dominates (the central reflectance is R_{max}= 19,426%). It can be expected that with the ageing of the black Electroink itself the minimal changes in reflectance of ΔR=1,647% appear.

5. Conclusion

The investigation results show that the ageing of the non coated printing substrates causes the decrease of the printed halftone dots. Printing on the aged paper results in the increased reproduction of the screen elements (especially in darker tones, 30% halftone value). For ROS Indigo electrophotographic laser head the generating of the screen elements with different sizes is characteristic. Their deviation

(20% halftone value) depends on the type of the printing substrate and on the way of ageing. In the combination, the non aged paper with the aged ink / aged paper with the non aged ink lead to the decrease of the standard deviation on natural papers (ΔSD_{off}=-9,41, SD_{rec}=-14,42μm), while the standard deviation (ΔSD_{gloss}=3,48 μm) increase on fine art paper. In the combination aged paper and aged ink/ aged paper and the non aged ink, on all the printing substrates, the standard deviation increases (ΔSD_{off}=3,08μm, ΔSD_{rec}=4,85 μm, ΔSD_{gloss}=5,26 μm).The non aged print on the non aged paper results in highest reflectance from the center of the printed screen dot. In relation to the non aged print on offset paper, the ageing of the screen dot creates the maximal difference in reflectance ΔR= 9,125%, while in relation to the print made on aged paper the maximal difference in reflectance appears ΔR=1,312%. In relation to the non aged print on the recycled paper, the ageing of prints creates the difference in reflectance ΔR= 2,557%, while in the relation to the print made on aged paper the difference in reflectance is insignificant ΔR=0,276%. Acting of the accelerated ageing influences mostly the print on the recycled paper creating an uneven reflectance from the surface of the printed screen element. Because of its coating (gloss), the fine art paper behaves differently from the uncoated papers. The accelerated ageing of fine art paper gives greater and darker screen elements. In relation to the print made on the non aged paper, the ageing of the screen dot causes the increase ΔR=21,843%, while the printing on the aged fine art paper causes further increase of ΔR=20,195%.

The research results are very useful for explanation of the ageing mechanism of material and prints in digital printing in the function of evaluation of the durable materials and materials of new formulation.

References

[1] Majnaric I Bolanca I Bolanca Z and Milkovic M 2005 Condition in digital printing of packaging on the print quality *Proc. International Conference on Digital Production Printing and Industrial Applications (Amsterdam, Netherlands, 9-13 May 2005)* ed R Borrell (IS&T vol 3) pp 97-99

[2] Majnaric I Bolanca S and Golubovic K 2007 The influence of ElectroInk pigmentation on the quality of indirect digital printing, *Proc. of 11th International Conference on Printing, Design and Graphic Communication (Zadar, Croatia, 26-29 September 2007)* ed Z Bolanca (Zagreb: Faculty of Graphic Arts) pp 85-89

[3] Landa Bet 1988 Toner for use in composition for developing latent electrostatic images, method of making the same and liquid composition using improved toner, *Patent United States 4,794,651, 27*

[4] Leach R and Pierce R 1999 *The Printing Ink Manual,*(Dordrecht: Kluwer Academic Publishers, Netherlands)

[5] Maraval M and Flieder F 1993 The stability of printing inks *Restaurator* **141** 41-171

[6] Halinova B Babiakova D Brezova V Durovic M Novotna M and Belanyi F 2002 The stability of offset inks on paper upon ageing *Dyes and Pigments* **54** 173-188

[7] Bonham J S and Rolniczak B 2006 Accelerated ageing of copy papers containing recycled fibre *Appita* **59** 365-369

[8] El-Sakhawy M 2005 Effect of bleaching sequence on paper ageing *Polymer Degradation and Stability* **87** 419-423

[9] Johansson A and Lennholm H 2000 Influences of SO_2 ,NO_2 and O_3 on the ageing of paper investigated by in situ diffuse reflectance FTIR and time-resolved trace gas analysis *Applied Surface Science* **161** 163-169

[10] Johansson A 2000 *Air pollution and paper deterioration, causes and remedies* Dissertation Göteborg University Sweden

[11] Fjellström H Höbglund H and Paulsson M 2007 Light-inducet yellowing of mecanical and

chemimechanical pulp sheets-Influence of wood raw material, process and aging method *Nordic Pulp and Paper Research* **22** 117-123
[12] Mcgarry P F Schmidt J A and Heitner C 2004 Accelerated light exposure of wood-containing pulps. Part I: Effects of wavelength and intensity *Tappi Journal* **3** 18-24
[13] Carter H P Begin D and Grattan 2000 Migration of volatile compounds through stacked sheet of paper during acclerated ageing *Restaurator* **21** 77-84
[14] *ISO 5360:2006 Anaesthetic vaporizers-Agent-specific filling systems*
[15] *TAPPI T 1219sp-98, Storage of Paper Samples for Optical Measurements and Colour Matching*

AUTHOR INDEX